Disk-Based Algorithms for Big Data

Disk-Based Algorithms for Big Data

Christopher G. Healey

North Carolina State University
Raleigh, North Carolina

CRC Press
Taylor & Francis Group
Boca Raton   London   New York

CRC Press is an imprint of the
Taylor & Francis Group, an **informa** business

A CHAPMAN & HALL BOOK

CRC Press
Taylor & Francis Group
6000 Broken Sound Parkway NW, Suite 300
Boca Raton, FL 33487-2742

© 2017 by Taylor & Francis Group, LLC
CRC Press is an imprint of Taylor & Francis Group, an Informa business

No claim to original U.S. Government works

Printed and bound in India by Replika Press Pvt. Ltd.

Printed on acid-free paper
Version Date: 20160916

International Standard Book Number-13: 978-1-138-19618-6 (Hardback)

Visit the Taylor & Francis Web site at
http://www.taylorandfrancis.com

and the CRC Press Web site at
http://www.crcpress.com

To Michelle

To my sister, the artist

To my parents

And especially, to D Belle and K2

Contents

List of Tables

List of Figures

Preface

This book is a product of recent advances in the areas of "big data," data analytics, and the underlying file systems and data management algorithms needed to support the storage and analysis of massive data collections.

We have offered an *Advanced File Structures* course for senior undergraduate and graduate students for many years. Until recently, it focused on a detailed exploration of advanced in-memory searching and sorting techniques, followed by an extension of these foundations to disk-based mergesort, B-trees, and extendible hashing.

About ten years ago, new file systems, algorithms, and query languages like the Google and Hadoop file systems (GFS/HDFS), MapReduce, and Hive were introduced. These were followed by database technologies like Neo4j, MongoDB, Cassandra, and Presto that are designed for new types of large data collections. Given this renewed interest in disk-based data management and data analytics, I searched for a textbook that covered these topics from a theoretical perspective. I was unable to find an appropriate textbook, so I decided to rewrite the notes for the *Advanced File Structures* course to include new and emerging topics in large data storage and analytics. This textbook represents the current iteration of that effort.

The content included in this textbook was chosen based of a number of basic goals:

- provide theoretical explanations for new systems, techniques, and databases like GFS, HDFS, MapReduce, Cassandra, Neo4j, and MongoDB,

- preface the discussion of new techniques with clear explanations of traditional algorithms like mergesort, B-trees, and hashing that inspired them,

- explore the underlying foundations of different technologies, and demonstrate practical use cases to highlight where a given system or algorithm is well suited, and where it is not,

- investigate physical storage hardware like hard disk drives (HDDs), solid-state drives (SSDs), and magnetoresistive RAM (MRAM) to understand how these technologies function and how they could affect the complexity, performance, and capabilities of existing storage and analytics algorithms, and

- remain accessible to both senior-level undergraduate and graduate students.

To achieve these goals, topics are organized in a bottom-up manner. We begin with the physical components of hard disks and their impact on data management, since HDDs continue to be common in large data clusters. We examine how data is stored and retrieved through primary and secondary indices. We then review different in-memory sorting and searching algorithms to build a foundation for more sophisticated on-disk approaches.

Once this introductory material is presented, we move to traditional disk-based sorting and search techniques. This includes different types of on-disk mergesort, B-trees and their variants, and extendible hashing.

We then transition to more recent topics: advanced storage technologies like SSDs, holographic storage, and MRAM; distributed hash tables for peer-to-peer (P2P) storage; large file systems and query languages like ZFS, GFS/HDFS, Pig, Hive, Cassandra, and Presto; and NoSQL databases like Neo4j for graph structures and MongoDB for unstructured document data.

This textbook was not written in isolation. I want to thank my colleague and friend Alan Tharp, author of *File Organization and Processing*, a textbook that was used in our course for many years. I would also like to recognize Michael J. Folk, Bill Zoellick, and Greg Riccardi, authors of *File Structures*, a textbook that provided inspiration for a number of sections in my own notes. Finally, Rada Chirkova has used my notes as they evolved in her section of *Advanced File Structures*, providing additional testing in a classroom setting. Her feedback was invaluable for improving and extending the topics the textbook covers.

I hope instructors and students find this textbook useful and informative as a starting point for their own investigation of the exciting and fast-moving area of storage and algorithms for big data.

Christopher G. Healey
June 2016

CHAPTER 1

Physical Disk Storage

FIGURE 1.1   The interior of a hard disk drive showing two platters, read/write heads on an actuator arm, and controller hardware

MASS STORAGE for computer systems originally used magnetic tape to record information. Remington Rand, manufacturer of the Remington typewriter and the UNIVAC mainframe computer (and originally part of the Remington Arms company), built the first tape drive, the UNISERVO, as part of a UNIVAC system sold to the U.S. Census Bureau in 1951. The original tapes were 1,200 feet long and held 224KB of data, equivalent to approximately 20,000 punch cards. Although popular until just a few years ago due to their high storage capacity, tape drives are inherently linear in how they transfer data, making them inefficient for anything other than reading or writing large blocks of sequential data.

Hard disk drives (HDDs) were proposed as a solution to the need for random access secondary storage in real-time accounting systems. The original hard disk drive, the Model 350, was manufactured by IBM in 1956 as part of their IBM RAMAC

(Random Access Method of Accounting and Control) computer system. The first RAMAC was sold to Chrysler's Motor Parts division in 1957. It held 5MB of data on fifty 24-inch disks.

HDDs have continued to increase their capacity and lower their cost. A modern hard drive can hold 3TB or more of data, at a cost of about $130, or $0.043/GB. In spite of the emergence of other storage technologies (e.g., solid state flash memory), HDDs are still a primary method of storage for most desktop computers and server installations. HDDs continue to hold an advantage in capacity and cost per GB of storage.

1.1 PHYSICAL HARD DISK

Physical hard disk drives use one or more circular platters to store information (Figure 1.1). Each platter is coated with a thin ferromagnetic film. The direction of magnetization is used to represent binary 0s and 1s. When the drive is powered, the platters are constantly rotating, allowing fixed-position heads to read or write information as it passes underneath. The heads are mounted on an actuator arm that allows them to move back and forth over the platter. In this way, an HDD is logically divided in a number of different regions (Figure 1.2).

- **Platter.** A non-magnetic, circular storage surface, coated with a ferromagnetic film to record information. Normally both the top and the bottom of the platter are used to record information.

- **Track.** A single circular "slice" of information on a platter's surface.

- **Sector.** A uniform subsection of a track.

- **Cylinder.** A set of vertically overlapping tracks.

An HDD is normally built using a stack of platters. The tracks directly above and below one another on successive platters form a cylinder. Cylinders are important, because the data in a cylinder can be read in one rotation of the platters, without the need to "seek" (move) the read/write heads. Seeking is usually the most expensive operation on a hard drive, so reducing seeks will significant improve performance.

Sectors within a track are laid out using a similar strategy. If the time needed to process a sector allows n additional sectors to rotate underneath the disk's read/write heads, the disk's interleave factor is $1 : n$. Each *logical* sector is separated by n positions on the track, to allow consecutive sectors to be read one after another without any rotation delay. Most modern HDDs are fast enough to support a $1 : 1$ interleave factor.

1.2 CLUSTERS

An operating system (OS) file manager usually requires applications to bundle information into a single, indivisible collection of sectors called a cluster. A cluster is a

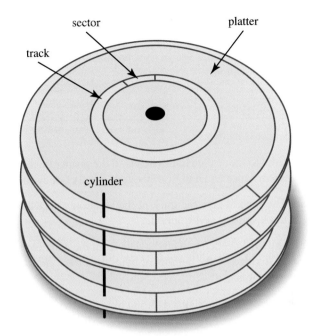

FIGURE 1.2 A hard disk drive's platters, tracks, sectors, and cylinders

contiguous group of sectors, allowing the data in a cluster to be read in a single seek. This is designed to improve efficiency.

An OS's file allocation table (FAT) binds the sectors to their parent clusters, allowing a cluster to be decomposed by the OS into a set of physical sector locations on the disk. The choice of cluster size (in sectors) is a tradeoff: larger clusters produce fewer seeks for a fixed amount of data, but at the cost of more space wasted, on average, within each cluster.

1.2.1 Block Allocation

Rather than using sectors, some OSs allowed users to store data in variable-sized "blocks." This meant users could avoid sector-spanning or sector fragmentation issues, where data either won't fit in a single sector, or is too small to fill a single sector. Each block holds one or more logical records, called the *blocking factor*. Block allocation often requires each block to be preceded by a count defining the block's size in bytes, and a key identifying the data it contains.

As with clusters, increasing the blocking factor can reduce overhead, but it can also dramatically increase track fragmentation. There are a number of disadvantages to block allocation.

- blocking requires an application and/or the OS to manage the data's organization on disk, and

- blocking may preclude the use of synchronization techniques supported by generic sector allocation.

1.3 ACCESS COST

The cost of a disk access includes

1. **Seek.** The time to move the HDD's heads to the proper track. On average, the head moves a distance equal to $\frac{1}{3}$ of the total number of cylinders on the disk.

2. **Rotation.** The time to spin the track to the location where the data starts. On average, a track spins $\frac{1}{2}$ a revolution.

3. **Transfer.** The time needed to read the data from the disk, equal to the number of bytes read divided by the number of bytes on a track times the time needed to rotate the disk once.

For example, suppose we have an 8,515,584 byte file divided into 16,632 sectors of size 512 bytes. Given a 4,608-byte cluster holding 9 sectors, we need a sequence of 1,848 clusters occupying at least 264 tracks, assuming a Barracuda HDD with sixty-three 512-byte sectors per track, or 7 clusters per track. Recall also the Barracuda has an 8 ms seek, 4 ms rotation delay, spins at 7200 rpm (120 revolutions per second), and holds 6 tracks per cylinder (Table 1.1).

In the best-case scenario, the data is stored contiguously on individual cylinders. If this is true, reading one track will load 63 sectors (9 sectors per cluster times 7 clusters per track). This involves a seek, a rotation delay, and a transfer of the entire track, which requires 20.3 ms (Table 1.2). We need to read 264 tracks total, but each cylinder holds 6 tracks, so the total transfer time is 20.3 ms per track times $\frac{264}{6}$ cylinders, or about 0.9 seconds.

TABLE 1.1 Specifications for a Seagate Barracuda 3TB hard disk drive

Property	Measurement
Platters	3
Heads	6
Rotation Speed	7200 rpm
Average Seek Delay	8 ms
Average Rotation Latency	4 ms
Bytes/Sector	512
Sectors/Track	63

TABLE 1.2 The estimated cost to access an 8.5GB file when data is stored "in sequence" in complete cylinders, or randomly in individual clusters

In Sequence	Random
A track (63 recs) needs: 1 seek + 1 rotation + 1 track xfer	A cluster (9 recs) needs: 1 seek + 1 rotation + $1/7$ track xfer
8 + 4 + 8.3 ms = 20.3 ms	8 + 4 + 1.1 ms = 13.1 ms
Total read:	Total read:
20.3 ms · $264/6$ cylinders = 893.2 ms	13.1 ms · 1848 clusters = 24208.8 ms

$$\times 27.1$$

In the worst-case scenario, the data is stored on clusters randomly scattered across the disk. Here, reading one cluster (9 sectors) needs a seek, a rotation delay, and a transfer of $1/7$ of a track, which requires 13.1 ms (Table 1.2). Reading all 1,848 clusters takes approximately 24.2 seconds, or about 27 times longer than the fully contiguous case.

Note that these numbers are, unfortunately, probably not entirely accurate. As larger HDDs have been offered, location information on the drive has switched from physical cylinder–head–sector (CHS) mapping to logical block addressing (LBA). CHS was 28-bits wide: 16 bits for the cylinder (0–65535), 4 bits for the head (0–15), and 8 bits for the sector (1–255), allowing a maximum drive size of about 128GB for standard 512-byte sectors.

LBA uses a single number to logically identify each block on a drive. The original 28-bit LBA scheme supported drives up to about 137GB. The current 48-bit LBA standard supports drives up to 144PB. LBA normally reports some standard values: 512 bytes per sector, 63 sectors per track, 16,383 tracks per cylinder, and 16 "virtual heads" per HDD. An HDD's firmware maps each LBA request into a physical cylinder, track, and sector value. The specifications for Seagate's Barracuda (Table 1.1) suggest it's reporting its properties assuming LBA.

1.4 LOGICAL TO PHYSICAL

When we write data to a disk, we start at a logical level, normally using a programming language's API to open a file and perform the write. This passes through to the OS, down to its file manager, and eventually out to the hard drive, where it's written as a sequence of bits represented as changes in magnetic direction on an HDD platter. For example, if we tried to write a single character P to the end of some file textfile, something similar to the following sequence of steps would occur.

1. Program asks OS through API to write P to the end of textfile.

2. OS passes request to file manager.

3. File manager looks up `textfile` in internal information tables to determine if the file is open for writing, if access restrictions are satisfied, and what physical file `textfile` represents.

4. File manager searches file allocation table (FAT) for physical location of sector to contain P.

5. File manager locates and loads last sector (or cluster containing last sector) into a system IO buffer in RAM, then writes P into the buffer.

6. File manager asks IO processor to write buffer back to proper physical location on disk.

7. IO processor formats buffer into proper-sized chunks for the disk, then waits for the disk to be available.

8. IO processor sends data to disk controller.

9. Disk controller seeks heads, waits for sector to rotate under heads, writes data to disk bit-by-bit.

File manager. The file manager is a component of the OS. It manages high-level file IO requests by applications, maintains information about open files (status, access restrictions, ownership), manages the FAT, and so on.

IO buffer. IO buffers are areas of RAM used to buffer data being read from and written to disk. Properly managed, IO buffers significantly improve IO efficiency.

IO processor. The IO processor is a specialized device used to assemble and disassemble groups of bytes being moved to and from an external storage device. The IO processor frees the CPU for other, more complicated processing.

Disk controller. The disk controller is a device used to manage the physical characteristics of an HDD: availability status, moving read/write heads, waiting for sectors to rotate under the heads, and reading and writing data on a bit level.

1.5 BUFFER MANAGEMENT

Various strategies can be used by the OS to manage IO buffers. For example, it is common to have numerous buffers allocated in RAM. This allows both the CPU and the IO subsystem to perform operations simultaneously. Without this, the CPU would be IO-bound. The pool of available buffers is normally managed with algorithms like LRU (least recently used) or MRU (most recently used).

Another option is known as locate mode. Here, the OS avoids copying buffers from program memory to system memory by (1) allowing the file manager to access program memory directly or (2) having the file manager provide an application with the locations of internal system buffers to use for IO operations.

A third approach is scatter–gather IO. Here, incoming data can be "scattered" among a collection of input buffers, and outgoing data can be "gathered" from a collection of output buffers. This avoids the need to explicitly reconstruct data into a single, large buffer.

CHAPTER 2

File Management

FIGURE 2.1   A typical data center, made up of racks of CPU and disk clusters

F ILES, IN their most basic form, are a collection of bytes. In order to manage
files efficiently, we often try to impose a structure to their contents by organizing
them in some logical manner.

2.1   LOGICAL COMPONENTS

At the simplest level, a file's contents can be broken down into a variety of logical
components.

- **Field.** A single (indivisible) data item.

- **Array.** A collection of equivalent fields.

- **Record.** A collection of different fields.

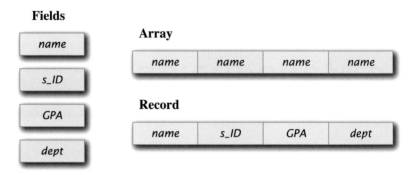

FIGURE 2.2   Examples of individual fields combined into an array (equivalent fields) and a record (different fields)

In this context, we can view a file as a stream of bytes representing one or more logical entities. Files can store anything, but for simplicity we'll start by assuming a collection of equivalent records.

2.1.1   Positioning Components

We cannot simply write data directly to a file. If we do, we lose the logical field and record distinctions. Consider the example below, where we write a record with two fields: last_name and first_name. If we write the values of the fields directly, we lose the separation between them. Ask yourself, "If we later needed to read last_name and first_name, how would a computer program determine where the last name ends and the first name begins?"

$$last\_name = Solomon$$
$$first\_name = Mark \implies SolomonMark$$

In order to manage fields in a file, we need to include information to identify where one field ends and the next one begins. In this case, you might use captial letters to mark field separators, but that would not work for names like O'Leary or MacAllen. There are four common methods to *delimit* fields in a file.

1. **Fixed length.** Fix the length of each field to a constant value.

2. **Length indicator.** Begin each field with a numeric value defining its length.

3. **Delimiter.** Separate each field with a delimiter character.

4. **Key–value pair.** Use a "keyword=value" representation to identify each field and its contents. A delimiter is also needed to separate key–value pairs.

Different methods have their own strengths and weaknesses. For example, fixed-length fields are easy to implement and efficient to manage, but they often provide

TABLE 2.1 Methods to logically organize data in a file: (a) methods to delimit fields; (b) methods to delimit records

Method	Advantages	Disadvantages
fixed length	simple, supports efficient access	may be too small, may waste space
length indicator	fields fit data	space needed for count
delimiter	fields fit data	space needed for delimiter, delimiter must be unique
key–value	efficient if many fields are empty	space needed for keyword, delimiter needed between keywords

(a)

Method	Advantages	Disadvantages
fixed length	simple, supports efficient access	may be too small, may waste space
field count	records fit fields	space required for count, variable length
length indicator	records fit fields	space needed for length, variable length
delimiter	records fit fields	space needed for delimiter, unique delimiter needed, variable length
external index	supports efficient access	indirection needed through index file

(b)

insufficient space if the field is too small, or wasted space if it's too large. Table 2.1a describes some advantages and disadvantages of each of the four methods.

Records have a similar requirement: the need to identify where each record starts and ends. Not surprisingly, methods to delimit records are similar to, but not entirely the same as, strategies to delimit fields.

1. **Fixed length.** Fix the length of each record to a constant value.

2. **Field count.** Begin each record with a numeric value defining the number of fields it holds.

3. **Length indicator.** Begin each record with a numeric value defining its length.

4. **Delimiter.** Separate each record with a delimiter character.

5. **External index.** Use an external index file to track the start location and length of each record.

Table 2.1b describes some advantages and disadvantages of each method for delimiting records. You don't need to use the same method to delimit fields and records. It's entirely possible, for example, to use a delimiter to separate fields within a record, and then to use an index file to locate each record in the file.

2.2 IDENTIFYING RECORDS

Once records are positioned in a file, a related question arises. When we're searching for a target record, how can we identify the record? That is, how can we distinguish the record we want from the other records in the file?

The normal way to identify records is to define a primary key for each record. This is a field (or a collection of fields) that uniquely identifies a record from all other possible records in a file. For example, a file of student records might use student ID as a primary key, since it's assumed that no two students will ever have the same student ID.

It's usually not a good idea to use a real data field as a key, since we cannot guarantee that two records won't have the same key value. For example, it's fairly obvious we wouldn't use last name as a primary key for student records. What about some combination of last name, middle name, and first name? Even though it's less likely, we still can't guarantee that two different students don't have the same last, middle, and first name. Another problem with using a real data field for the key value is that the field's value can change, forcing an expensive update to parts of the system that link to a record through its primary key.

A better approach is to generate a non-data field for each record as its added to the file. Since we control this process, we can guarantee each primary key is unique and immutable, that is, the key value will not change after it's initially defined. Your student ID is an example of this approach. A student ID is a non-data field, unique to each student, generated when a student first enrolls at the university, and never changed as long as a student's records are stored in the university's databases.[1]

2.2.1 Secondary Keys

We sometimes use a non-unique data field to define a *secondary key*. Secondary keys do not identify individual records. Instead, they subdivide records into logical groups with a common key value. For example, a student's major department is often used as a secondary key, allowing us to identify Computer Science majors, Industrial Engineering majors, Horticulture majors, and so on.

We define secondary keys with the assumption that the grouping they produce is commonly required. Using a secondary key allows us to structure the storage of records in a way that makes it computationally efficient to perform the grouping.

---

[1] Primary keys *usually* never change, but on rare occasions they must be modified, even when this forces an expensive database update. For example, student IDs at NC State University used to be a student's social security number. For obvious privacy reasons this was changed, providing every student with a new, system-generated student ID.

2.3 SEQUENTIAL ACCESS

Accessing a file occurs in two basic ways: sequential access, where each byte or element in a file is read one-by-one from the beginning of the file to the end, or direct access, where elements are read directly throughout the file, with no obvious systematic pattern of access.

Sequential access reads through in sequence from beginning to end. For example, if we're searching for patterns in a file with grep, we would perform sequential access.

This type of access supports sequential, or linear, search, where we hunt for a target record starting at the front of the file, and continue until we find the record or we reach the end of the file. In the best case the target is the first record, producing $O(1)$ search time. In the worst case the target is the last record, or the target is not in the file, producing $O(n)$ search time. On average, if the target is in the file, we need to examine about $n/2$ records to find the target, again producing $O(n)$ search time.

If linear search occurs on external storage—a file—versus internal storage—main memory—we can significantly improve *absolute* performance by reducing the number of seeks we perform. This is because seeks are much more expensive than in-memory comparisons or data transfers. In fact, for many algorithms we'll equate performance to the number of seeks we perform, and not on any computation we do after the data has been read into main memory.

For example, suppose we perform record blocking during an on-disk linear search by reading m records into memory, searching them, discarding them, reading the next block of m records, and so on. Assuming it only takes one seek to locate each record block, we can potentially reduce the worst-case number of seeks from n to n/m, resulting in a significant time savings. Understand, however, that this only reduces the absolute time needed to search the file. It does not change search efficiency, which is still $O(n)$ in the average and worst cases.

In spite of its poor efficiency, linear search can be acceptable in certain cases.

1. Searching files for patterns.

2. Searching a file with only a few records.

3. Managing a file that rarely needs to be searched.

4. Performing a secondary key search on a file where many matches are expected.

The key tradeoff here is the cost of searching versus the cost of building and maintaining a file or data structure that supports efficient searching. If we don't search very often, or if we perform searches that require us to examine most or all of the file, supporting more efficient search strategies may not be worthwhile.

2.3.1 Improvements

If we know something about the types of searches we're likely to perform, it's possible to try to improve the performance of a linear search. These strategies are known as *self-organizing*, since they reorganize the order of records in a file in ways that could make future searches faster.

Move to Front. In the move to front approach, whenever we find a target record, we move it to the front of the file or array it's stored in. Over time, this should move common records near the front of the file, ensuring they will be found more quickly. For example, if searching for one particular record was very common, that record's search time would reduce to $O(1)$, while the search time for all the other records would only increase by at most one additional operation. Move to front is similar to an LRU (least recently used) paging algorithm used in an OS to store and manage memory or IO buffers.[2]

The main disadvantage of move to front is the cost of reorganizing the file by pushing all of the preceding records back one position to make room for the record that's being moved. A linked list or indexed file implementation can ease this cost.

Transpose. The transpose strategy is similar to move to front. Rather than moving a target record to the front of the file, however, it simply swaps it with the record that precedes it. This has a number of possible advantages. First, it makes the reorganization cost much smaller. Second, since it moves records more slowly toward the front of the file, it is more stable. Large "mistakes" do not occur when we search for an uncommon record. With move to front, whether a record is common or not, it always jumps to the front of the file when we search for it.

Count. A final approach assigns a count to each record, initially set to zero. Whenever we search for a record, we increment its count, and move the record forward past all preceding records with a lower count. This keeps records in a file sorted by their search count, and therefore reduces the cost of finding common records.

There are two disadvantages to the count strategy. First, extra space is needed in each record to hold its search count. Second, reorganization can be very expensive, since we need to do actual count comparisons record-by-record within a file to find the target record's new position. Since records are maintained in sorted search count order, the position can be found in $O(\lg n)$ time.

2.4 DIRECT ACCESS

Rather than reading through an entire file from start to end, we might prefer to jump directly to the location of a target record, then read its contents. This is efficient, since the time required to read a record reduces to a constant $O(1)$ cost. To perform direct access on a file of records, we must know where the target record resides. In other words, we need a way to convert a target record's key into its location.

One example of direct access you will immediately recognize is array indexing. An array is a collection of elements with an identical type. The index of an array

---

[2]Move to front is similar to LRU because we push, or discard, the least recently used records toward the end of the array.

TABLE 2.2 A comparison of average case linear search performance versus worst case binary search performance for collections of size n ranging from 4 records to 2^{64} records

Method				n		
	4	16	256	65536	4294967296	2^{64}
Linear	2	8	128	32768	2147483648	2^{63}
Binary	2	4	8	16	32	64
Speedup	1×	2×	16×	2048×	67108864×	$2^{57}×$

element is its key, and this "key" can be directly converted into a memory offset for the given element. In C, this works as follows.

```
int a[ 256 ]
a[ 128 ] ≡ *(a + 128)
```

This is equivalent to the following.

```
a[ 128 ] ≡ &a + ( 128 * sizeof( int ) )
```

Suppose we wanted to perform an analogous direct-access strategy for records in a file. First, we need fixed-length records, since we need to know how far to offset from the front of the file to find the i-th record. Second, we need some way to convert a record's key into an offset location. Each of these requirements is non-trivial to provide, and both will be topics for further discussion.

2.4.1 Binary Search

As an initial example of one solution to the direct access problem, suppose we have a collection of fixed-length records, and we store them in a file sorted by key. We can find target records using a binary search to improve search efficiency from $O(n)$ to $O(\lg n)$.

To find a target record with key k_t, we start by comparing against key k for the record in the middle of the file. If $k = k_t$, we retrieve the record and return it. If $k > k_t$, the target record could only exist in the lower half of the file—that is, in the part of the file with keys smaller than k—so we recursively continue our binary search there. If $k < k_t$ we recursively search the upper half of the file. We continue cutting the size of the search space in half until the target record is found, or until our search space is empty, which means the target record is not in the file.

Any algorithm that discards half the records from consideration at each step needs at most $\log_2 n = \lg n$ steps to terminate, in other words, it runs in $O(\lg n)$ time. This is the key advantage of binary search versus an $O(n)$ linear search. Table 2.2 shows some examples of average case linear search performance versus worst case binary search performance for a range of collection sizes *n*.

Unfortunately, there are also a number of disadvantages to adopting a binary search strategy for files of records.

1. The file must be sorted, and maintaining this property is very expensive.

2. Records must be fixed length, otherwise we cannot jump directly to the *i*-th record in the file.

3. Binary search still requires more than one or two seeks to find a record, even on moderately sized files.

Is it worth incurring these costs? If a file was unlikely to change after it is created, and we often need to search the file, it might be appropriate to incur the overhead of building a sorted file to obtain the benefit of significantly faster searching. This is a classic tradeoff between the initial cost of construction versus the savings after construction.

Another possible solution might be to read the file into memory, then sort it prior to processing search requests. This assumes that the cost of an in-memory sort whenever the file is opened is cheaper than the cost of maintaining the file in sorted order on disk. Unfortunately, even if this is true, it would only work for small files that can fit entirely in main memory.

2.5 FILE MANAGEMENT

Files are not static. In most cases, their contents change over their lifetime. This leads us to ask, "How can we deal efficiently with additions, updates, and deletions to data stored in a file?"

Addition is straightforward, since we can store new data either at the first position in a file large enough to hold the data, or at the end of the file if no suitable space is available. Updates can also be made simple if we view them as a deletion followed by an addition.

Adopting this view of changes to a file, our only concern is how to efficiently handle record deletion.

2.5.1 Record Deletion

Deleting records imposes a number of requirements on a file management system. Records normally need to be marked for deletion, since they may not be immediately removed from the file. Whether the records are fixed length or variable length can also complicate the deletion strategy. Indeed, we will conclude that, because fixed-length records are so advantageous, if a file holds variable-length records, then we will construct a secondary *index* file with fixed length entries to allow us to manage the original file in an efficient manner.

Storage Compaction. One very simple deletion strategy is to delete a record, then—either immediately or in the future—compact the file to reclaim the space used by the record.

This highlights the need to recognize which records in a file have been deleted. One option is to place a special "deleted" marker at the front of the record, and change the file processing operations to recognize and ignore deleted records.

It's possible to delay compacting until convenient, for example, until after the user has finished working with the file, or until enough deletions have occurred to warrant compacting. Then, all the deletions in the file can be compacted in a single pass. Even in this situation, however, compacting can be very expensive. Moreover, files that must provide a high level of availability (e.g., a credit card database) may never encounter a "convenient" opportunity to compact themselves.

2.5.2 Fixed-Length Deletion

Another strategy is to dynamically reclaim space when we add new records to a file. To do this, we need ways to

- mark a record as being deleted, and

- *rapidly* find space previously used by deleted records, so that this space can be reallocated to new records added to the file.

As with storage compaction, something as simple as a special marker can be used to tag a record as deleted. The space previously occupied by the deleted record is often referred to as a *hole*.

To meet the second requirement, we can maintain a stack of holes (deleted records), representing a stack of available spaces that should be reclaimed during the addition of new records. This works because *any* hole can be used to hold a new record when all the records are the same, fixed length.

It's important to recognize that the hole stack must be *persistent*, that is, it must be maintained each time the file is closed, or recreated each time the file is opened. One possibility is to write the stack directly in the file itself. To do this, we maintain an offset to the location of the first hole in the file. Each time we delete a record, we

- mark the record as deleted, creating a new hole in the file,

- store within the new hole the current head-of-stack offset, that is, the offset to the *next* hole in the file, and

- update the head-of-stack offset to point to the offset of this new hole.

When a new record is added, if holes exist, we grab the first hole, update the head-of-stack offset based on its next hole offset, then reuse its space to hold the new record. If no holes are available, we append the new record to the end of the file.

Figure 2.3 shows an example of adding four records with keys A, B, C, and D to a file, deleting two records B and D, then adding three more records X, Y, and Z. The following steps occur during these operations.

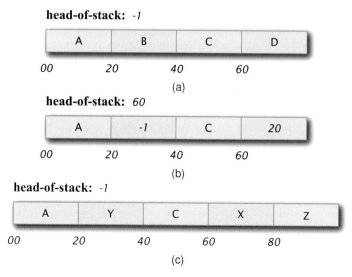

FIGURE 2.3 Fixed-length record deletion: (a) A, B, C, and D are added to a file; (b) B and D are deleted; (c) X, Y, and Z are added

1. The head-of-stack offset is set to −1, since an empty file has no holes.

2. A, B, C, and D are added. Since no holes are available (the head-of-stack offset is −1), all four records are appended to the end of the file (Figure 2.3a).

3. B is deleted. Its next hole offset is set to −1 (the head-of-stack offset), and the head-of-stack is set to 20 (B's offset).

4. D is deleted. Its next hole offset is set to 20, and the head-of-stack is updated to 60 (Figure 2.3b).

5. X is added. It's placed at 60 (the head-of-stack offset), and the head-of-stack offset is set to 20 (the next hole offset).

6. Y is added at offset 20, and the head-of-stack offset is set to −1.

7. Z is added. Since the head-of-stack offset is −1, it's appended to the end of the file (Figure 2.3c).

To adopt this in-place deletion strategy, a record must be large enough to hold the deleted marker plus a file offset. Also, the head-of-stack offset needs to be stored within the file. For example, the head-of-stack offset could be appended to the end of the file when it's closed, and re-read when it's opened.

2.5.3 Variable-Length Deletion

A more complicated problem is supporting deletion and dynamic space reclamation when records are variable length. The main issue is that new records we add may not exactly fit the space occupied by previously deleted records. Because of this, we need to (1) find a hole that's big enough to hold the new record; and (2) determine what to do with any leftover space if the hole is larger than the new record.

The steps used to perform the deletion are similar to fixed-length records, although their details are different.

- mark a record as being deleted, and

- add the hole to an availability list.

The availability list is similar to the stack for fixed-length records, but it stores both the hole's offset and its size. Record size is simple to obtain, since it's normally part of a variable-length record file.

First Fit. When we add a new record, how should we search the availability list for an appropriate hole to reallocate? The simplest approach walks through the list until it finds a hole big enough to hold the new record. This is known as the first fit strategy.

Often, the size of the hole is larger than the new record being added. One way to handle this is to increase the size of the new record to exactly fit the hole by padding it with extra space. This reduces external fragmentation—wasted space between records—but increases internal fragmentation—wasted space within a record. Since the entire purpose of variable-length records is to avoid internal fragmentation, this seems like a counterproductive idea.

Another approach is to break the hole into two pieces: one exactly big enough to hold the new record, and the remainder that forms a new hole placed back on the availability list. This can quickly lead to significant external fragmentation, however, where the availability list contains many small holes that are unlikely to be big enough to hold any new records.

In order to remove these small holes, we can try to merge physically adjacent holes into new, larger chunks. This would reduce external fragmentation. Unfortunately, the availability list is normally not ordered by physical location, so performing this operation can be expensive.

Best Fit. Another option is to try a different placement strategy that makes better use of the available holes. Suppose we maintain the availability list in ascending order of hole size. Now, a first fit approach will always find the smallest hole capable of holding a new record. This is called best fit. The intuition behind this approach is to leave the smallest possible chunk on each addition, minimizing the amount of space wasted in a file due to external fragmentation.

Although best fit can reduce wasted space, it incurs an additional cost to maintain the availability list in sorted order. It can also lead to a higher cost to find space for newly added records. The small holes created on each addition are put at the front of the availability list. We must walk over all of these holes when we're searching for a location for a new record. Finally, by their nature, the small holes created on addition will often never be big enough to hold a new record, and over time they can add up to a significant amount of wasted space.

Worst Fit. Suppose we instead kept the availability list sorted in descending order of hole size, with the largest available hole always at the front of the list. A first fit strategy will now find the largest hole capable of storing a new record. This is called worst fit. The idea here is to create the largest possible remaining chunk when we split a hole to hold a new record, since larger chunks are more likely to be big enough for a new record at some point in the future.

Worst fit also reduces the search for an appropriate hole to an O(1) operation. If the first hole on the availability list is big enough to hold the new record, we use it. Otherwise, none of the holes on the availability list will be large enough for the new record, and we can immediately append it to the end of the file.

2.6 FILE INDEXING

Our previous discussions suggest that there are many advantages to fixed-length records in terms of searching and file management, but they impose serious drawbacks in terms of efficient use of space. File indexing is an attempt to separate the issue of storage from the issue of access and management through the use of a secondary *file index*. This allow us to use a fixed-length *index file* for managing and searching for records and a variable-length *data file* to store the actual contents of the records.

2.6.1 Simple Indices

We begin with a simple index: an array of key–offset pairs that *index* or report the location of a record with key k at offset position p in the data file. Later, we will look at more complicated data structures like B-trees and external hash tables to manage indices more efficiently.

A key feature of an index is its support for indirection. We can rearrange records in the data file simply by rearranging the indices that references the records, without ever needing to touch the data file itself. This can often be more efficient; moreover, features like pinned records—records that are not allowed to change their position— are easily supported. Indices also allow us to access variable-length records through a fixed-length index, so we can support direct access to any record through its index entry's offset.

As an example, consider a data file with variable-length records describing music files. The primary key for these records is a combination of the distribution company's label and the recording's ID number. Table 2.3a shows entries from the index file, consisting of a key–offset pair. In this example, the size of each key is fixed to the maximum possible key length. This introduces some internal fragmentation in the index file, but this is acceptable relative to the ability to avoid fragmentation in the data file and still support efficient search.

TABLE 2.3 Indexing variable-length records: (a) index file; (b) data file

key	offset
ANG3795	152
COL31809	338
COL38358	196
DGI39201	382
DGI8807	241
. . .	

(a)

offset	record
0	LON | 2312 | Romeo & Juliet | ...
62	RCA | 2626 | Quartet in C# | ...
117	WAR | 2369 | Touchstone | ...
152	ANG | 3795 | Symphony #9 | ...
196	COL | 38358 | Nebraska | ...
. . .	

(b)

It's also possible to store other (fixed-length) information in the index file. For example, since we are working with variable-length records, we might want to store each record's length together with its key–offset pair.

2.6.2 Index Management

When an index file is used, changes to the data file require one or more corresponding updates to the index file. Normally, the index is stored as a separate file, loaded when the data file is opened, and written back, possibly in a new state, when the data file is closed.

For simple indices, we assume that the entire index fits in main memory. Of course, for large data files this is often not possible. We will look at more sophisticated approaches that maintain the index on disk later in the course.

Addition. When we add a new record to the data file, we either append it to the end of the file or insert it in an internal hole, if deletions are being tracked and if an appropriately sized hole exists. In either case, we also add the new record's key and offset to the index. Each entry must be inserted in sorted key order, so shifting elements to open a space in the index, or resorting of the index, may be necessary. Since the index is held in main memory, this will be much less expensive than a disk-based reordering.

Deletion. Any of the deletion schemes previously described can be used to remove a record from the data file. The record's key–offset pair must also be found and removed from the index file. Again, although this may not be particularly efficient, it is done in main memory.

Notice that the use of an index solves the issue of pinned records—records that are not allowed to change their position in the data file. A record that is deleted from the data file is no longer present in the index. Any subsequent reordering is done only on the index, so record positions in the data file never change. Any other data structures referring to pinned records—for example, the availability list—don't need to worry that the location of a deleted record might be moved to some new offset in the file.

Update. Updating either changes a record's primary key, or it does not. In the latter case, nothing changes in the index. In the former case, the key in the record's key–offset pair must be updated, and the index entry may need to be moved to maintain a proper sorted key order. It's also possible in either case to handle an update with a deletion followed by an add.

2.6.3 Large Index Files

None of the operations on an index prevents us from storing it on disk rather than in memory. Performance will decrease dramatically if we do this, however. Multiple seeks will be needed to locate keys, even if we use a binary search. Reordering the index during addition or deletion will be prohibitively expensive. In these cases, we will most often switch to a different data structure to support indexing, for example, B-trees or external hash tables.

There is another significant advantage to simple indices. Not only can we index on a primary key, we can also index on secondary keys. This means we can provide multiple pathways, each optimized for a specific kind of search, into a single data file.

2.6.4 Secondary Key Index

Recall that secondary keys are not unique to each record; instead, they partition the records into groups or classes. When we build a secondary key index, its "offset" references are normally into the primary key index and not into the data file itself. This buffers the secondary index and minimizes the number of entries that need to be updated when the data file is modified.

Consider Table 2.4, which uses composer as a secondary key to partition music files by their composer. Notice that the reference for each entry is a primary key. To retrieve records by the composer Beethoven, we would first retrieve all primary keys for entries with the secondary key Beethoven, then use the primary key index to locate and retrieve the actual data records corresponding to Beethoven's music files.

Note that the secondary index is sorted first by secondary key, and within that by primary key reference. This order is required to support combination searches.

Addition. Secondary index addition is very similar to primary key index addition: a new entry must be added, in sorted order, to each secondary index.

Deletion. As with the primary key index, record deletion would normally require every secondary index to remove its reference to the deleted record and close any hole this creates in the index. Although this is done in main memory, it can still be expensive.

If the secondary index referenced the data file directly, there would be few alternatives to this approach. The space in the data file is normally marked as available

TABLE 2.4 A secondary key index on composer

secondary key	primary key reference
Beethoven	ANG3795
Beethoven	DGI39201
Beethoven	DGI8807
Beethoven	RCA2626
Corea	WAR23699
Dvorak	COL31809
. . .	

and subsequently reused, so if the secondary index were not updated, it would end up referencing space that pointed to new, out-of-date information.

If the secondary index references the primary key index, however, it is possible to simply remove the reference from the primary index, then stop. Any request for deleted records through the secondary index will generate a search on the primary index that fails. This informs the secondary index that the record it's searching for no longer exists in the data file.

If this approach is adopted, at some point in the future the secondary index will need to be "cleaned up" by removing all entries that reference non-existent primary key values.

Update. During update, since the secondary index references through the primary key index, a certain amount of buffering is provided. There are three "types" of updates that a secondary index needs to consider.

1. The secondary key value is updated. When this happens, the secondary key index must update its corresponding entry, and possibly move it to maintain sorted order.

2. The primary key value is updated. In this situation, the secondary index must be searched to find the old primary key value and replace it with the new one. A small amount of sorting may also be necessary to maintain the proper by-reference within-key ordering.

3. Neither the primary nor the secondary key values are updated. Here, no changes are required to the secondary key index.

Boolean Logic. Secondary key indices can be used to perform Boolean logic searches. This is done by retrieving "hit lists" from searches on different secondary

TABLE 2.5 Boolean and applied to secondary key indices for composer and symphony to locate recordings of Beethoven's Symphony #9

Beethoven	Symphony #9	Result
ANG3795 ——	ANG3795 ——	ANG3795
DGI39201	COL31809	
DGI8807 ——	DGI8807 ——	DGI8807
RCA2626		

indices, then merging the results using Boolean operators to produce a final list of records to retrieve.

For example, suppose we wanted to find all recordings of Beethoven's Symphony #9. Assuming we had secondary key indices for composer and symphony, we could use these indices to search for recordings by Beethoven, and symphonies named Symphony #9, then and these two lists to identify the target records we want (Table 2.5).

Notice this is why secondary indices must be sorted by secondary key, and *within* key by primary key reference. This allows us to rapidly walk the hit lists and merge common primary keys. Other types of Boolean logic like or, not, and so on can also be applied.

2.6.5 Secondary Key Index Improvements

There are a number of issues related to secondary key indices stored as simple arrays:

- the array must be reordered for each addition, and

- the secondary key value is duplicated in many cases, wasting space.

If we use a different type of data structure to store the secondary key index, we can address these issues.

Inverted List. One possible alternative is to structure each entry in the secondary index as a secondary key value and n primary key reference slots (Table 2.6a). Each time a new record with a given secondary key is added, we add the record's primary key, in sorted order, to the set of references currently associated with that secondary key.

The advantage of this approach is that we only need to do a local reorganization of a (small) list of references when a new record is added. We also only need to store each secondary key value once.

A disadvantage of an inverted list is deciding how big to make n, the maximum number of references attached to each secondary key. This is the standard problem

TABLE 2.6 Alternative secondary index data structures: (a) an inverted list; (b, c) a secondary key list of values and head-of-list offsets into a linked list of references

Key	References
Beethoven	ANG3795 DGI39201 DGI8807 RCA2626
Prokofiev	ANG36193 LON2312

(a)

Key	Offset
Beethoven	30
Corea	70
Dvorak	90

(b)

Offset	Primary Key	Next
0	LON2312	∅
10	RCA2626	∅
20	WAR23699	∅
30	ANG3795	80
40	COL38358	∅
50	DGI8807	10
60	MER75016	∅
70	COL31809	∅
80	DGI39201	50

(c)

of choosing an array size: if it's too small, the secondary key will fail due to lack of space, but if it's too large, significant amounts of space may be wasted.

Linked List. An obvious solution to the size issue of inverted lists is to switch from an array of primary key references to a dynamically allocated linked list of references. Normally, we don't create separate linked lists for each secondary key. Instead, we build a single reference list file referenced by the secondary key index (Table 2.6b, c).[3]

Now, each secondary key entry holds an offset to the first primary key in the secondary key's reference list. That primary key entry holds an offset to the next primary key in the reference list, and so on. This is similar to the availability list for deleted records in a data file.

There are a number of potential advantages to this linked list approach:

- we only need to update the secondary key index when a record is added,[4] or when a record's secondary key value is updated,

---

[3] If memory is available, it's possible to read the reference list into an internal data structure.

[4] We assume new records are added to the front of a secondary key list.

- the primary key reference list is entry sequenced, so it never needs to be sorted, and

- the primary key reference list uses fixed-length records, so it is easy to implement deletion, and we can also use direct access to jump to any record in the list.

There are also some possible disadvantages to a linked list of references. The main problem involves the locality of the entries that make up a secondary key's reference list. When the reference list is stored on disk, each primary key we retrieve will often require a disk seek to jump to the location in the list that holds that key. This means reconstructing a reference list with m primary keys could involve m seeks, an expense we often can't afford to incur.

CHAPTER 3

Sorting

FIGURE 3.1   Frozen raspberries are sorted by size prior to packaging

S ORTING IS one of the fundamental operations will we study in this course. The need to sort data has been critical since the inception of computer science. For example, bubble sort, one of the original sorting algorithms, was analyzed to have average and worst case performance of $O(n^2)$ in 1956. Although we know that comparison sorts theoretically cannot perform better than $O(n \lg n)$ in the best case, better performance is often possible—although not guaranteed—on real-world data. Because of this, new sorting algorithms continue to be proposed.

We start with an overview of sorting collections of records that can be stored entirely in memory. These approaches form the foundation for sorting very large data collections that must remain on disk.

3.1   HEAPSORT

Heapsort is another common sorting algorithm. Unlike Quicksort, which runs in $O(n \lg n)$ in the best and average cases, but in $O(n^2)$ in the worse case, heapsort

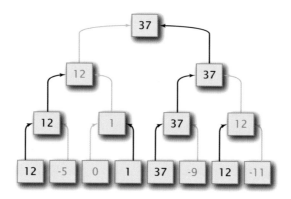

FIGURE 3.2 A tournament sort used to identify the largest number in the initial collection

guarantees $O(n \lg n)$ performance, even in the worst case. In absolute terms, however, heapsort is slower than Quicksort. If the possibility of a worst case $O(n^2)$ is not acceptable, heapsort would normally be chosen over Quicksort.

Heapsort works in a manner similar to a tournament sort. In a tournament sort all pairs of values are compared and a "winner" is promoted to the next level of the tournament. Successive winners are compared and promoted until a single overall winner is found. Tournament sorting a set of numbers where the bigger number wins identifies the largest number in the collection (Figure 3.2). Once a winner is found, we reevaluate its winning path to promote a second winner, then a third winner, and so on until all the numbers are promoted, returning the collection in reverse sorted order.

It takes $O(n)$ time to build the initial tournament structure and promote the first element. Reevaluating a winning path requires $O(\lg n)$ time, since the height of the tournament tree is $\lg n$. Promoting all n values therefore requires $O(n \lg n)$ time. The main drawback of tournament sort is that it needs about $2n$ space to sort a collection of size n.

Heapsort can sort in place in the original array. To begin, we define a heap as an array $A[1 \dots n]$ that satisfies the following rule.[1]

$$\left. \begin{array}{l} A[i] \geq A[2i] \\ A[i] \geq A[2i+1] \end{array} \right\} \quad \text{if they exist} \tag{3.1}$$

To sort in place, heapsort splits A into two parts: a heap at the front of A, and a partially sorted list at the end of A. As elements are promoted to the front of the heap, they are swapped with the element at the end of the heap. This grows the sorted list and shrinks the heap until the entire collection is sorted. Specifically, heapsort executes the following steps.

---

[1] Note that heaps are indexed starting at 1, not at 0 like a C array. The heapsort algorithms will not work unless the array starts at index 1.

1. Manipulate A into a heap.

2. Swap $A[1]$—the largest element in A—with $A[n]$, creating a heap with $n-1$ elements and a partially sorted list with 1 element.

3. Readjust $A[1]$ as needed to ensure $A[1 \ldots n-1]$ satisfy the heap property.

4. Swap $A[1]$—the second largest element in A—with $A[n-1]$, creating a heap with $n-2$ elements and a partially sorted list with 2 elements.

5. Continue readjusting and swapping until the heap is empty and the partially sorted list contains all n elements in A.

We first describe how to perform the third step: readjusting A to ensure it satisfies the heap property. Since we started with a valid heap, the only element that might be out of place is $A[1]$. The following sift algorithm pushes an element $A[i]$ at position i into a valid position, while ensuring no other elements are moved in ways that violate the heap property (Figure 3.3b,c).

sift(A, i, n)
Input: $A[\ ]$, heap to correct; i, element possibly out of position; n, size of heap

```
while  i ≤ ⌊n/2⌋ do
    j = i*2                                          // j = 2i
    k = j+1                                          // k = 2i + 1

    if  k ≤ n and A[k] ≥ A[j]  then
    |    lg = k                                      // A[k] exists and A[k] ≥ A[j]
    else
    |    lg = j                                      // A[k] doesn't exist or A[j] > A[k]
    end

    if  A[i] ≥ A[lg]  then
    |    return                                      // A[i] ≥ larger of A[j], A[k]
    end

    swap A[i], A[lg]
    i = lg
end
```

So, to move $A[1]$ into place after swapping, we would call sift(A, 1, n-1). Notice that sift isn't specific to $A[1]$. It can be used to move any element into place. This allows us to use sift to convert A from its initial configuration into a heap.

heapify(A, n)
Input: $A[\ ]$, array to heapify; n, size of array

```
i = ⌊n/2⌋
while i ≥ 1 do
|    sift( A, i, n )          // Sift A[i] to satisfy heap constraints
|    i--
end
```

The heapify function assumes the rightmost element that might violate the heap constraints is $A[\lfloor n/2 \rfloor]$. This is because elements past $\lfloor n/2 \rfloor$ have nothing to compare against, so by default $A[\lfloor n/2 \rfloor + 1 \ldots n]$ satisfy the heap property (Figure 3.3b).

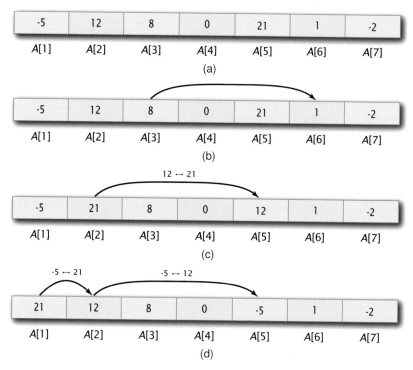

FIGURE 3.3 Heap construction: (a) A in its initial configuration; (b) sifting $A[3]$ to confirm its position; (c) sifting $A[2] = 12$ into place by swapping with $A[5] = 21$; (d) sifting $A[1] = -5$ into place by swapping with $A[2] = 21$, then with $A[5] = 12$

After $A[\lfloor n/2 \rfloor]$ is "in place," everything from $A[\lfloor n/2 \rfloor ...n]$ satisfies the heap property. We move back to the element at $A[\lfloor n/2 \rfloor - 1]$ and sift it into place. This continues until we sift $A[1]$ into place. At this point, A is a heap.

We use the heapify and sift functions to heapsort an array A as follows.

```
heapsort(A, n)
```
Input: $A[\ ]$, array to sort; n, size of array

```
heapify( A, n )                          // Convert A into a heap
i = n
while i ≥ 2 do
    swap A[ 1 ], A[ i ]    // Move largest heap element to sorted list
    i--

    sift( A, 1, i )          // Sift A[1] to satisfy heap constraints
end
```

Performance. Unlike Quicksort, heapsort has identical best case, worst case, and

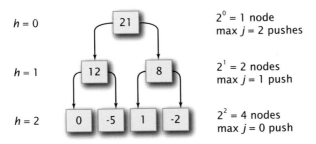

$h = 0$

21

$2^0 = 1$ node
max $j = 2$ pushes

$h = 1$

12   8

$2^1 = 2$ nodes
max $j = 1$ push

$h = 2$

0   -5   1   -2

$2^2 = 4$ nodes
max $j = 0$ push

FIGURE 3.4   A heap represented as a binary tree; each node's value is greater than or equal to the values in all of its child nodes

average case performance.[2,3] Given this, we choose to explain a worst case scenario. We start with sift. The maximum number of swaps we could see would involve moving $A[1]$ to $A[2]$, then to $A[4]$, and so on up to $A[\lfloor n/2 \rfloor]$. Walking through $\lfloor n/2 \rfloor$ elements in jumps of size 1, 2, 4 ... requires O($\lg n$) steps.

The heapsort function first converts A into a heap, then applies sift $n - 1$ times. The swap–sift loop therefore requires O($n \lg n$) time to execute.

What about heapify? It calls sift $\lfloor n/2 \rfloor$ times, so it needs no more than O($n \lg n$) time to run. We could stop here, since this is equivalent to the time needed to sort the resulting heap. It turns out, however, that O($n \lg n$) is not tight. heapify actually runs in O(n) time.

To understand this, you need to count the maximum number of jumps each element $A[1] \ldots A[\lfloor n/2 \rfloor]$ would need to make to be pushed into place. Consider a heap drawn as a tree (Figure 3.4). In this configuration, the heap property guarantees the value at any node is greater than or equal to the values of all of its children.

Given a heap forming a complete tree of height h with $n = 2^{h+1} - 1$ elements, the 2^h leaf elements are not pushed, the 2^{h-1} elements at level $h - 1$ can be pushed at most one time, the 2^{h-2} elements at level $h - 2$ can be pushed at most two times, and so on. In general, the 2^{h-j} elements at level $h - j$ can be pushed at most j times. The maximum number of sift pushes $P(h)$ in a heap tree of height h is therefore

$$P(h) = \sum_{j=1}^{h} j \, 2^{h-j} = \sum_{j=1}^{h} j \frac{2^h}{2^j} = 2^h \sum_{j=1}^{h} \frac{j}{2^j} \tag{3.2}$$

How do we solve $\sum_{j=1}^{h} j/2^j$? First, we can equate this to $\sum_{j=0}^{h} j/2^j$, since $j/2^j = 0$ for $j = 0$. Next, consider the infinite geometric sum

$$\sum_{j=0}^{\infty} x^j = \frac{1}{(1-x)} \tag{3.3}$$

---

[2] The analysis of heapsort. Schaffer and Sedgewick. *Journal of Algorithms 15*, 1, 76–100, 1993.
[3] On the best case of heapsort. Bollobás, Fenner and Frieze. *Journal of Algorithms 20*, 2, 205–217, 1996.

If we differentiate both sides and multiply by x, we obtain

$$\sum_{j=0}^{\infty} \left(x^j\right)' x = \left(\frac{1}{(1-x)}\right)' x$$

$$\sum_{j=0}^{\infty} \left(jx^{j-1}\right) x = \left(\frac{1}{(1-x)^2}\right) x \qquad (3.4)$$

$$\sum_{j=0}^{\infty} jx^j = \frac{x}{(1-x)^2}$$

If we replace x with $\frac{1}{2}$, we obtain the function we want on the left hand side of the equation, $\sum_{j=0}^{\infty} j/2^j$. Substituting $x = \frac{1}{2}$ on the right hand side produces 2, so

$$\sum_{j=0}^{\infty} \frac{j}{2^j} = 2 \qquad (3.5)$$

Our sum runs to h and not ∞, so we're bounded above by this equation.

$$P(n) = 2^h \sum_{j=1}^{h} j/2^j \qquad (3.6)$$
$$\leq 2^h \, 2 = 2^{h+1}$$

Since $n = 2^{h+1} - 1$, this simplifies to $P(n) \leq n + 1 = O(n)$. For completeness, since $\lfloor n/2 \rfloor$ elements must be considered for a push, $P(n)$ is also bounded below by $\Omega(n)$. Since $P(n)$ is bounded above by $O(n)$ and below by $\Omega(n)$, it has running time $\Theta(n)$.

3.2 MERGESORT

Mergesort was one of the original sorting algorithms, proposed by John von Neumann in 1945. Similar to Quicksort, it works in a divide and conquer manner to sort an array A of size n.

1. Divide A into n/l sorted sublists, or *runs*, of length l.

2. *Merge* pairs of runs into new runs of size $2l$.

3. Continue merging runs until $l = n$, producing a sorted version of A.

In its most basic implementation we initially use $l \leq 1$, since a run of size zero or one is, by default, sorted. Runs are merged to produce new runs of size $2, 4 \ldots$ up to a final run of size n (Figure 3.5).

The bulk of the work is done in the merge step. This operation is simple: given two sorted runs, we walk through the runs in sequence, choosing the smaller of the

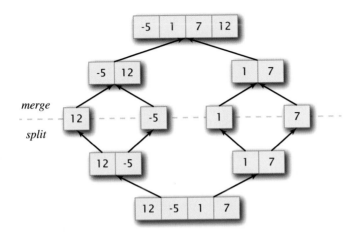

FIGURE 3.5 An array *A* split into runs of length 1, then recursively merged into a sorted result

two values to output to a new, merged run. Assume the runs are contiguous in *A*, with the left run occupying *A*[*lf* ... *mid* − 1] and the right run occupying *A*[*mid* ... *rt*].

```
merge(A, lf, mid, rt)
```
Input: *A*[ ], runs to merge; *lf*, left run start; *mid*, right run start; *rt*, right run end

```
i = lf                          // Current element in left run
j = mid                         // Current element in right run

B = [ ]
k = 0

while k ≤ rt - lf + 1 do
    if i < mid and (j > rt or A[i] ≤ A[j]) then
        | B[k++] = A[i++]        // Left run element exist, is smallest
    else
        | B[k++] = A[j++]        // Right run element exists, is smallest
    end
end
copy B to A[lf ... rt]          // Copy merged runs
```

```
msort(A, beg, end)
```
Input: *A*[ ], array to split/merge; *beg*, start of subarray; *end*, end of subarray

```
if end - beg ≥ 1 then
    mid = ⌊ ( beg + end ) / 2 ⌋         // Start of right run

    msort( A, beg, mid - 1 )     // Recursively create, sort left run
    msort( A, mid, end )         // Recursively create, sort right run
    merge( A, beg, mid, end )               // Merge sorted runs
end
```

The recursive function `msort` uses `merge` to split A into sorted runs, then merge the runs together to produce a sorted result. To sort A, we call `msort(A, 0, n-1)`.

Performance. In our implementation, best, average, and worst case performance for mergesort are all $O(n \lg n)$.

First, consider the performance of `merge`. It walks through both runs, so it needs $O(m)$ time, where $m = rt - lf + 1$ is the combined size of the runs.

Next, consider the recursive function `msort`. Similar to Quicksort, `msort` divides an array A of size n into two subarrays of size $n/2$, then four subarrays of size $n/4$, and so on down to n subarrays of size 1. The number of divisions needed is $\lg n$, and the total amount of work performed to merge pairs of runs at each level of the recursion is $O(n)$. Mergesort therefore runs in $O(n \lg n)$ time.

Mergesort also requires $O(n)$ additional space (the B array in the `merge` function) to hold merged runs.

3.3 TIMSORT

Timsort was proposed by Tim Peters in 2002. It was initially implemented as a standard sorting method in Python. It is now being offered as a built-in sorting method in environments like Android and Java. Timsort is a hybrid sorting algorithm, a combination of insertion sort and an adaptive mergesort, built specifically to work well on real-world data.

Timsort revolves around the idea that an array A is a sequence of sorted runs: *ascending* runs where $A[i] \leq A[i + 1] \leq A[i + 2]$ ... and *descending* runs where $A[i] > A[i + 1] > A[i + 2]$ ....[4] Timsort leverages this fact by merging the runs together to sort A.

Every run will be at least 2 elements long,[5] but if A is random, very long runs are unlikely to exist. Timsort walks through the array, checking the length of each run it finds. If the run is too short, Timsort extends its length, then uses insertion sort to push the addition elements into sorted order.

How short is "too short?" Timsort defines a minimum run length *minrun*, based on the size of A.[6] This guarantees that no run is less than *minrun* long, and for a random A, almost all the runs will be exactly *minrun* long, which leads to a very efficient mergesort.

As runs are identified or created, their starting position and length are stored on a run stack. Whenever a new run is added to the stack, a check is made to see whether any runs should be merged. Suppose that X, Y, and Z are the last three runs added to the top of the run stack. Only consecutive runs can be merged, so the two options are to create $(X + Y) Z$ or $X (Y + Z)$.

Deciding when to merge is a balance between maintaining runs to possibly exploit good merges as new runs are found, versus merging quickly to exploit memory

---

[4]Timsort reverses descending runs in place, converting all runs to ascending.

[5]A run starting with the last element in A will only be 1 element long.

[6]*minrun* is selected from the range $32 \ldots 65$ | $n/minrun = 2^x$ (i.e., a power of 2), or when this is not possible, | *minrun* $\approx 2^x$ and *minrun* $< 2^x$ (i.e., close to, but strictly less than a power of 2).

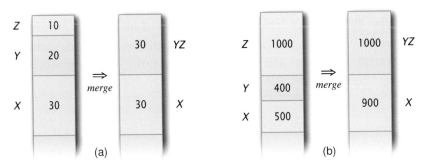

FIGURE 3.6 Managing the Timsort run stack: (a) $X \le Y + Z$ and $Z < X$ so merge Y, Z; (b) $X \le Y + Z$ and $X < Z$ so merge X, Y

caching and to avoid a run stack that uses large amounts of memory. To do this, Timsort enforces two constraints on the lengths of the last three runs on the run stack.

1. $X > Y + Z$

2. $Y > Z$

If $X \le Y + Z$, the smaller of X and Z is merged with Y, with ties favoring Z. Figure 3.6 shows two examples of merging Y with Z and X with Y. Notice that in both cases the second constraint $Y > Z$ is still violated, so we would continue merging the last three runs on the top of the stack until the constraints were satisfied, or until there is only one run on the stack.

Merging. Although it's possible to apply a standard mergesort to merge two ascending runs X and Y, Timsort tries to be smarter to improve absolute performance. Merging starts in the standard way, comparing $X[0]$ to $Y[0]$ and moving the smaller of the two to an output buffer. Timsort calls this *one pair at a time* mode. In addition to walking through the runs, we maintain a count c of how many times in a row the winning element comes from the same run.

Galloping. If c reaches a threshold *min-gallop*, we enter *galloping mode*. Now, we take element $X[0]$ at the top of X and search Y directly for the position p where it belongs. We copy $Y[0 \ldots p - 1]$ to the output buffer, followed by $X[0]$. Then we take $Y[0]$ at the top of Y and search X for the position p where *it* belongs, copying $X[0 \ldots p - 1]$, then $Y[0]$ to the output buffer. We continue galloping until both searches of X and Y copy subarrays that have less than *min-gallop* elements in them. At this point, we switch back to one pair at a time mode.

Searching. To find $X[0]$ in Y, we compare in turn to $Y[0]$, $Y[1]$, $Y[3]$, $Y[7]$, … $Y[2^j - 1]$, searching for $k \mid Y[2^{k-1} - 1] < X[0] \le Y[2^k - 1]$. At this point we know that $X[0]$ is somewhere in the $2^{k-1} - 1$ elements from $Y[2^{k-1} - 1]$ to $Y[2^k - 1]$. A regular binary search is used on this range to find the final position for $X[0]$. The time needed to find k is $\approx \lg |Y|$. Some extra time is also needed to perform a binary search on the subarray containing $X[0]$'s position.

TABLE 3.1   Performance for optimized versions of mergesort, Quicksort, and Timsort on input data that is identical, sequential, partially sequential, random without duplicates, and random with sequential steps; numbers indicate speedup versus mergesort

Algorithm	identical	sequential	part seq	random	part rand
mergesort	1.0×	1.0×	1.0×	1.0×	1.0×
Quicksort	6.52×	6.53×	**1.81×**	**1.25×**	1.53×
Timsort	**6.87×**	**6.86×**	1.28×	0.87×	**1.6×**

If we perform a binary search directly on Y to position $X[0]$, it takes $\lceil \lg|Y| \rceil$ comparisons, regardless of where $X[0]$ lies in Y. This means that straight binary search only wins if the subarray identified using an adaptive search is large. It turns out that, if the data in A is random, $X[0]$ usually occurs near the front of Y, so long subarrays are extremely rare.[7] Even if long subarrays do occur, galloping still finds and copies them in $O(\lg n)$ versus $O(n)$ for standard mergesort, producing a huge time savings.

Performance. In theoretical terms, Timsort's best case performance is $O(n)$, and its average and worst case performances are $O(n \lg n)$. However, since Timsort is tuned to certain kinds of real-world data—specifically, partially sorted data—it's also useful to compare absolute performance for different types of input.

Table 3.1 shows absolute sort time speedups for Quicksort and Timsort versus mergesort for different input types.[8] Here, Timsort performed well for data that was sequential or partially random, while Quicksort performed best for data that was fully random. This suggests that Timsort has overall performance comparable to Quicksort, and if data is often sorted or nearly sorted, Timsort may outperform Quicksort.

---

[7]Optimistic sorting and information theoretic complexity. McIlroy. *Proceedings of the 4th Annual ACM-SIAM Symposium on Discrete Algorithms (SODA 1993)*, Austin, TX, pp. 467–474, 1993.

[8]http://blog.quibb.org/2009/10/sorting-algorithm-shootout.

CHAPTER 4

Searching

FIGURE 4.1   A web browser search field

S EARCHING IS the second fundamental operation we will study in this course. As with sorting, efficient searching is a critical foundation in computer science. We review $O(n)$ linear search and $O(\lg n)$ binary search, then discuss more sophisticated approaches. Two of these techniques, trees and hashing, form the basis for searching very large data collections that must remain on disk.

4.1   LINEAR SEARCH

The simplest search takes a collection of n records and scans through them from start to end, looking for a record with a target key k_t.

Best case performance—when the target is the first record—is $O(1)$. Worst case performance—when the target is the last record or the target is not in the collection—is $O(n)$. On average, we assume we must search about $n/2$ records to find a target contained in the collection, which also runs in $O(n)$ time.

The main purposes of linear search are twofold. First, since it is very simple to implement, we sometimes use linear search when n is small or when searching is rare. Second, linear search represents a hard upper bound on search performance. If a search algorithm requires $O(n)$ time (or more), we'd often be better off using a simple linear search.

4.2 BINARY SEARCH

If a collection is maintained in sorted order, we can perform a binary search.

```
binary_search(k, arr, l, r)
Input:  k, target key; arr, sorted array to search; l, left endpoint; r, right endpoint

n = r-l+1
if n ≤ 0 then
|   return -1                                  // Searching empty range
end

c = l + ⌊n / 2⌋
if k == arr[c] then
|   return c                                   // Target record found
else if k < arr[c] then
|   return binary_search( k, arr, l, c-1 )     // Search left half
else
|   return binary_search( k, arr, c+1, r )     // Search right half
end
```

Calling `binary_search( k_t, arr, 0, n-1 )` initiates a search. This compares the target key k_t to the key at the center of the collection k_c. If $k_t = k_c$, the target record is found. Otherwise, sorted order tells us if $k_t < k_c$ then k_t is left of the center record, otherwise $k_t > k_c$ and k_t is right of the center record. Searching continues recursively until k_t is found, or until the collection is exhausted.

Binary search discards half the collection ($^n/_2$ records) on its first comparison, then half the remaining collection ($^n/_4$ records) on its next comparison, and so on. Any operation that halves the size of the collection on each step runs in $O(\lg n)$ time.

4.3 BINARY SEARCH TREE

If we choose to implement binary search, we must decide what type of data structure to use to manage the sorted collection. One possibility is a sorted array. As shown above, this provides $O(\lg n)$ search performance. Unfortunately, *maintaining* the collection is not as fast. Inserting a new record requires $O(\lg n)$ time to find its correct position, but then requires $O(n)$ time to shift part of the collection to make space to hold the new record. Deletion similarly requires $O(n)$ time to fill the hole left by the old record. There is the also the practical issue of choosing a good initial array size, and the need to allocate more space if the array overflows.

A common alternative is a binary search tree, or BST. A BST is a *tree* structure made up of nodes, each of which holds a record and references to two (possibly empty) child subtrees (Figure 4.2). The subtrees are normally labelled left and right. Each node in the BST satisfies the following ordering properties.

1. All records in a node's left subtree have keys smaller than the node's key.

2. All records in a node's right subtree have keys larger than the node's key.

Given this ordering, performing a binary search with a BST is very simple.

```
bst_search(k, node)
Input: k, target key; node, node to search

if node == null then
|   return null                            // Searching empty tree
end

if k == node.key then
|   return node                            // Target record found
else if k < node.key then
|   return bst_search( k, node.left )      // Search left subtree
else
|   return bst_search( k, node.right )     // Search right subtree
end
```

The logic applied here is identical to binary search, since BSTs are designed specifically to support this search strategy.

Insertion. To insert a record with key k_t into a BST, we search for k_t in the tree. When we reach an empty subtree, we insert a new node containing k_t's record. Since insertion requires a search followed by a constant time operation, insertion performance is identical to search performance.

Deletion. To delete a record with key k_t from a BST, we search for k_t in the tree. If a node containing k_t is found, we remove it and correct the BST based on three possible configuration cases.

1. If the node has no children, nothing needs to be done (Figure 4.2a).

2. If the node has one subtree, promote its subtree's root (Figure 4.2b).

3. If the node has two subtrees (Figure 4.2c)

    (a) Find the successor to k_t—the smallest value greater than k_t—in the right subtree by walking right once, then walking left as far as possible.

    (b) Remove the successor from the tree; since it has an empty left subtree it must match Case 1 or Case 2 above.

    (c) Promote the successor to the node's position.

Again, since deletion requires a search followed by a constant time operation, deletion performance is identical to search performance.

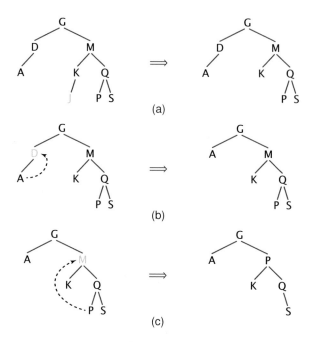

FIGURE 4.2 Deletion from a BST: (a) deleting J, a node with no subtrees; (b) deleting D, a node with one subtree; (c) deleting M, a node with two subtrees

Performance. Search performance in a BST depends on its shape. Suppose the BST is balanced: for any node in the tree, the height of its left and right subtrees are about equal. For example, the left BST in Figure 4.2a is roughly balanced, since the difference in left and right subtree heights is no more than 1 throughout the tree. A balanced BST with n records has a height of about $\lg n$, producing best case search performance of $O(\lg n)$ time.

A fully unbalanced BST is one in which every internal node has one subtree empty. Here, the BST degenerates into a linked list of n nodes, producing worst case search performance of $O(n)$. Unfortunately, the common situation of inserting records with keys in sorted or nearly sorted order produces this worst case.

4.4 k-d TREE

A k-dimensional or k-d tree is a binary tree used to subdivide a collection of records into *ranges* for k different attributes in each record. The k-d tree was proposed by Jon Louis Bentley in 1975 to support associative, or multiattribute, searches.[1] For example, we could take a collection of weather reports and divide them by properties like latitude, longitude, temperature, or precipitation. We could then make queries like "Return all records with temperature $< 0°C$ and precipitation > 4 cm."

---

[1] Multidimensional binary search trees used for associative searching. Bentley. *Communications of the ACM 18*, 9, 509–517, 1975.

TABLE 4.1 Estimated *ht* and *wt* of Snow White and the Seven Dwarfs

Name	ht	wt
Sleepy	36	48
Happy	34	52
Doc	38	51
Dopey	37	54
Grumpy	32	55
Sneezy	35	46
Bashful	33	50
Ms. White	65	98

k-d trees are often used as a method of flexible secondary indexing, although there is no reason why primary keys cannot participate as one of the k dimensions.

A k-d tree's structure is similar to a BST, except at each level of the tree we rotate between the k dimensions used to subdivide the tree's records. For example, a 2-d tree using attributes temperature and pressure would subdivide based on temperature at the root node, subdivide based on pressure in the root node's children, based again on temperature in the children's children, and so on.

4.4.1 k-d Tree Index

Like any binary tree, each k-d tree node contains a key value k_c and two subtrees: left and right. Unlike a BST, however, records are normally not stored in the internal nodes. Instead, the target key k_t is used to choose which subtree to enter: the left subtree if $k_t \leq k_c$, or the right subtree if $k_t > k_c$. Leaf nodes contain collections of records, specifically all records that satisfy the conditions along the root-to-leaf path.

Suppose we wanted to use information about Snow White and the Seven Dwarfs to build a k-d tree index. We will use the attributes height (*ht*) and weight (*wt*) as the two dimensions to subdivide records in the tree.

The construction algorithm works identically to BST, except that we rotate between the $k = 2$ dimensions as we walk through each level of the tree.

1. Sleepy is inserted into the root of the tree, which uses *ht* as its subdivision attribute.

2. Happy and Doc are inserted as children of Sleepy. Since Happy's $ht = 34 \leq 36$, Happy goes to the left of the root. Doc's $ht = 38 > 36$, so he goes to the right of the root (Figure 4.3a). Both Happy and Doc use *wt* as their subdivision attribute.

3. Dopey is inserted next. His $ht = 37$ puts him to the right of the root, and his $wt = 51$ puts him to the left of his parent (Figure 4.3b).

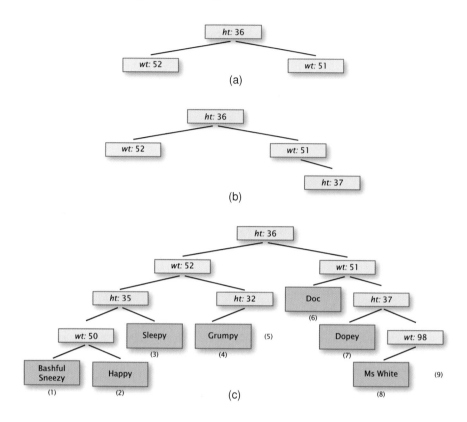

FIGURE 4.3 A k-d tree indexed by *ht* and *wt* of Snow White and the Seven Dwarfs: (a) the first three insertions, *ht* subdivides the root, *wt* subdivides the second level; (b) an *ht* subdivision node on the third level; (c) the final tree

4. The remaining dwarfs and Snow White are inserted using an identical approach.

Once the k-d tree index is complete, it acts as a method to locate records based on their *ht* and *wt* attributes. Buckets are placed at each null subtree, ready to hold additional entries as they are inserted. Figure 4.3c shows the buckets containing the initial dwarfs and Snow White.

Interpretation. A k-d tree index subdivides the *k*-dimensional space of all possible records into subspaces over a continuous range of values for each dimension. Another way to visualize a k-d tree index is as a subdivision of *k*-dimensional space using $(k - 1)$-dimensional cutting planes that represent each entry in the index.

The height–weight index in Figure 4.3 can be visualized this way. Since the index uses $k = 2$ dimensions, we subdivide a 2D plane using 1D lines into regions that represent each bucket in the tree (Figure 4.4).

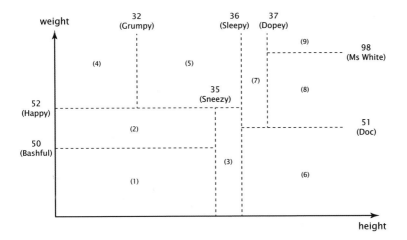

FIGURE 4.4 A subdivision of the $k = 2$ dimensional plane into subspaces representing each bucket in the k-d tree

4.4.2 Search

To search for records that match attribute ranges in a k-d tree, we perform the following operations.

1. Identify all paths whose internal nodes satisfy the target attribute ranges. This may produce multiple paths.

2. Perform an in-memory search of each path's bucket for records that match the target criteria.

For example, suppose we search for records with $ht \leq 36$ and $wt \leq 47$.

- at the root, branch left ($ht \leq 36$),

- at the next node, branch left again ($wt \leq 49$),

- at the next node, branch left and right ($ht \leq 35$ and $ht > 35$ both fall within the target range of $ht \leq 36$),

- along the right path we reach bucket 3, and

- along the left path, branch left ($wt \leq 50$), reaching bucket 1.

The search produces two paths that identify buckets 1 and 3 as potentially containing target records. Examining either Figure 4.3c or Figure 4.4 shows that

> Bucket 1: $ht \leq 35$ and $wt \leq 50$
> Bucket 3: $35 < ht \leq 36$ and $wt \leq 52$

Both buckets may include records with $ht \leq 36$ and $wt \leq 47$. Moreover, no other buckets in the table could contain these types of records.

4.4.3 Performance

It should be clear that a k-d tree's index has a critical impact on its performance. Ideally, the index should subdivide data stored in the tree in a balanced way, for example, by placing all the buckets at the same level in the tree, and by storing about the same number of elements in each bucket. If the data is known a priori, median elements can be used to construct the index.[2]

Our k-d tree is an example of an index that is designed for a certain class of individuals: those with $ht \leq 37$ and $wt \leq 55$. If we try to store a large number of records outside this range, they will all be forced into only one or two different buckets.

For dynamic trees, maintaining balance in the index is more complicated. Here, adaptive k-d trees can be used to try to adjust the index when buckets become too full or out of balance. A simple, although potentially inefficient, suggestion is to take all the records in an out-of-balance area of the tree, then repartition them and reconstruct the affected region of the index.[3]

4.5 HASHING

A second major class of algorithms used for efficient searching are hash algorithms. A hash function converts a key k_t into a numeric value h on a fixed range $0 \ldots n - 1$. h is used as a location or an address for k_t within a hash table A of size n. This is analogous to indexing on an array, since we can store and retrieve k_t at $A[h]$. If the hash function runs in constant time, search, insertion, and deletion are $O(1)$ operations.

Unfortunately, the number of possible records $m \gg n$ is normally much larger than the number of hash values n. Given this, three important properties distinguish hashing from using h to directly index into A.

1. The hash value for k_t should appear random.

2. Hash values should be distributed uniformly over the range $0 \ldots n - 1$.

3. Two different keys k_s and k_t can hash to the same h, producing a *collision*.

4.5.1 Collisions

Collisions are a major issue, particularly if each location in a hash table can only hold one record. If two records both hash to the same location, what should we do?

One answer might be, "Choose a hash function that doesn't produce collisions." This is harder than it sounds, however. Suppose we're storing credit card information, and we decide to use the credit card number as a key. For card numbers of the form 0000 0000 0000 0000, there are $m = 10^{16}$ possible numbers (10 quadrillion).

---

[2] An algorithm for finding best matches in logarithmic expected time. Friedman, Bentley, and Finkel. *ACM Transactions on Mathematical Software 3*, 3, 209–226, 1977.

[3] *Foundations of Multidimensional and Metric Data Structures*. Samet. Morgan Kaufmann, San Francisco, 2006.

Clearly, it's not possible to create an in-memory array of size $n = 10^{16}$. Of course, every possible card number isn't being used, in part because the credit card companies haven't issued 10^{16} cards, and in part because different parts of a credit card number are dependent in various ways[4] (e.g., certain parts of the card number represent checksums to ensure the card is valid, other parts define card type, bank number, and so on). Card numbers do span a reasonable part of the range from around 1×10^{15} to 9×10^{15}, however, so an array is still not feasible.

Even if the total number and range of the keys is small, it's still difficult to define a perfect hashing function with no collisions. For example, if we wanted to store $m = 4000$ keys in an array of size $n = 5000$, it's estimated that only 1 in 10^{120000} functions will be perfect. Given this, a more tractable approach is to reduce the number of collisions and to determine how to handle collisions when they occur.

4.5.2 Hash Functions

Here is a common fold-and-add hash function.

1. Convert k_t to a numeric sequence.

2. Fold and add the numbers, checking for overflow.

3. Divide the result by a prime number and return the remainder as h.

Consider k_t = Subramanian. We convert this into a numeric sequence by mapping each character to its ASCII code, then binding pairs of ASCII codes.

S u	b r	a m	a n	i a	n
85 117	98 114	97 109	97 110	105 97	110

Assume the largest character pair is zz with combined ASCII codes of 122122. To manage overflow during addition, we divide by prime number 125299 slightly larger than this maximum after each add, and keep the remainder.

$$
\begin{aligned}
85117 + 98114 &= 193231 \bmod 125299 = 67932 \\
67932 + 97109 &= 165041 \bmod 125299 = 39742 \\
35742 + 97110 &= 136852 \bmod 125299 = 11553 \qquad (4.1) \\
11553 + 10597 &= 22150 \bmod 125299 = 22150 \\
22150 + 110 &= 22260 \bmod 125299 = 22260
\end{aligned}
$$

We divide the result of 22260 by the size of the hash table, which itself should be prime. Here, we assume A has size $n = 101$, producing a final h of

$$h = 22260 \bmod 101 = 40 \qquad (4.2)$$

---

[4] http://www.mint.com/blog/trends/credit-card-code-01202011

Other useful hash functions exist. For example, we could convert k_t to a numeric sequence, square the sequence, and use the middle digits modulo the hash table size for h. Or we could convert the numeric sequence to a different base and use the converted value modulo the hash table size for h.

4.5.3 Hash Value Distributions

Given a hash table size of n used to hold r records, what is the likelihood that

1. No key hashes to a particular address in the table,

2. One key hashes to a particular address,

3. Two keys hash to a particular address,

and so on? Assume our hash function uniformly distributes its hash values. For any single key the probability it hashes to a given address is b, and the probability that it doesn't hash to that address (i.e., it hashes to some other address) is a.

$$b = \frac{1}{n}, \quad a = 1 - \frac{1}{n} \tag{4.3}$$

Given a and b, suppose we insert two keys into the hash table. We can compute individual cases, for example, the probability that the first key "hits" an address and the second key "misses," or the probability that both keys hit.

$$
\begin{aligned}
b\,a &= \frac{1}{n}\,(1 - \frac{1}{n}) = \frac{1}{n} - \frac{1}{n^2} \\
b\,b &= \frac{1}{n}\frac{1}{n} \quad = \frac{1}{n^2}
\end{aligned} \tag{4.4}
$$

What is the probability that x of r keys hash to a common address? First, we need to determine how many ways there are to arrange x hits in a sequence of r keys. This is the binomial coefficient, or choose probability r choose x.

$$C = \binom{r}{x} = \frac{r!}{x!\,(r - x)!} \tag{4.5}$$

Given C, the probability of x hits in r keys at a common address is

$$C\,b^x\,a^{r-x} = C\left(\frac{1}{n}\right)^x \left(1 - \frac{1}{n}\right)^{r-x} \tag{4.6}$$

Because of the $r!$ in its equation, C is expensive to compute. Fortunately, the Poisson distribution $\Pr(x)$ does a good job of estimating our probability.

$$C\,b^x\,a^{r-x} \approx \Pr(x) = \frac{\left(\frac{r}{n}\right)^x e^{-\left(\frac{r}{n}\right)}}{x!} \tag{4.7}$$

Since x is normally small, the $x!$ in the denominator is not an issue. Consider an extreme case, where we want to store $r = 1000$ keys in a hash table of size $n = 1000$.

Here, $r/_n = 1$. We can use this ratio to calculate $Pr(0)$, the probability an address is empty, $Pr(1)$, the probability one key hashes to an address, $Pr(2)$, the probability two keys hash to an address, and so on.

$$Pr(0) = \frac{1^0\, e^{-1}}{0!} = 0.368$$

$$Pr(1) = \frac{1^1\, e^{-1}}{1!} = 0.368 \tag{4.8}$$

$$Pr(2) = \frac{1^2\, e^{-1}}{2!} = 0.184$$

Based on these probabilities, and given our hash table size of $n = 1000$, we expect about $n\,Pr(0) = 1000 \cdot 0.368 = 368$ entries that are empty, $n\,Pr(1) = 368$ entries holding 1 key, $n\,Pr(2) = 184$ entries that try to hold 2 keys, and so on.

4.5.4   Estimating Collisions

Consider our previous example with $r = n = 1000$. How many collisions do we expect to see in this situation? To answer this, we use the following hash table breakdown.

n Pr(0) = 368 entries in the table hold no keys, and $n\,Pr(1) = 368$ entries hold exactly 1 key. This means $1000 - n\,Pr(0) - n\,Pr(1) = 264$ entries try to hold more than 1 key. For each of these 264 entries, the first key is stored, and any subsequent key collides. A total of $368 + 264 = 632$ keys are stored, so $1000 - 632 = 368$ keys collide, for a collision rate of 36.8%. The collision rate does not depend on r or n alone, it depends on the packing density $r/_n$ of the hash table.

One way to reduce the collision rate is to increase the size of the hash table. Suppose we increase the table size to $n = 2000$, which halves the packing density to $r/_n = {}^{1000}/_{2000} = 0.5$.

$$Pr(0) = \frac{0.5^0\, e^{-0.5}}{0!} = 0.607$$

$$Pr(1) = \frac{0.5^1\, e^{-0.5}}{1!} = 0.304 \tag{4.9}$$

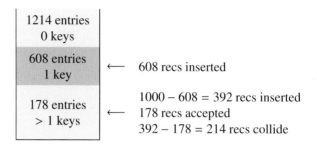

TABLE 4.2 Collision rates for different hash table densities

Density	10%	20	30	40	50	60	70	80	90	100
Collision	4.8%	9.4	13.6	17.6	21.4	24.8	28.1	31.2	34.1	36.8

Now $n \Pr(0) = 1214$ entries in the table hold no keys, and $n \Pr(1) = 608$ entries hold exactly 1 key. $1000 - n \Pr(0) - n \Pr(1) = 178$ entries try to hold more than 1 key. $1000 - 608 = 392$ keys hash to these positions, of which 178 are stored and $392 - 178 = 214$ collide, for a collision rate of 21.4%.

4.5.5 Managing Collisions

Table 4.2 shows that, even for very low packing densities, some collisions will still occur. Because of this, we need ways to manage a collision when it happens. We look at two common approaches: progressive overflow and multirecord buckets.

4.5.6 Progressive Overflow

One simple way to handle a collision on insertion is to hash a record's key, and if the resulting address h is already occupied, to walk forward through the table until an empty position is found.

To delete a record, we find and remove it. We also mark its position as *dirty* to remember that, although this position is empty, it was previously occupied.

```
progressive_insert(rec, tbl, n)
Input: rec, record to insert; tbl, hash table; n, table size

num = 0                                    // Number of insertion attempts
h = hash( rec.key )

while num < n do
    if tbl[ h ] is empty then
        tbl[ h ] = rec                     // Store record
        break
    else
        h = ( h + 1 ) % n                  // Try next table position
        num++
    end
end
```

 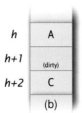

FIGURE 4.5 Progressive overflow: (a) records A, B, and C hash to *h*, forming a run; (b) B's position must be marked as dirty on removal to maintain C's insertion run

```
progressive_delete(key, tbl, dirty, n)
Input: key, key to remove; tbl, hash table; dirty, dirty entry table; n, table size

h = progressive_search( key, tbl, dirty, n )

if h != false then
     tbl[ h ] = empty                    // Set table position empty
     dirty[ h ] = true                   // Mark table position dirty
end


progressive_search(key, tbl, dirty, n)
Input: key, key to find; tbl, hash table; dirty, dirty entry table; n, table size

num = 0                                  // Number of compare attempts
h = hash( key )

while num < n do
     if key == tbl[ h ].key then
         return tbl[ h ]                       // Target record found
     else if tbl[ h ] is empty and !dirty[ h ] then
         return false                          // Search failed
     else
         h = ( h + 1 ) % n                 // Try next table position
         num++
     end
end

return false                             // Search failed
```

To search for a record, we hash its key to get *h*, then search from position *h* forward. If we find the record, the search succeeds. If we search the entire table without finding the record, the search fails. If we find an empty position whose dirty bit isn't set, the search also fails.

Why does the search stop at empty positions that aren't dirty, but jump over empty positions that are dirty? Suppose we insert three records A, B, and C that all hash to the same position *h*. A and B form a *run* in the table, a block of records that C must step over to insert itself (Figure 4.5a).

Next, we delete B, then search for C. The run that forced C to position $h + 2$ is gone (Figure 4.5b). The search algorithm wants to follow C's insertion path to find it. If we stopped at any empty entry, we would fail to find C. Marking position $h + 1$ as dirty tells the search algorithm, "Although this position is empty, it *may* have been part of a run when C was inserted, so keep searching."

Progressive overflow is simple to understand and implement, but it has a number of serious disadvantages.

1. The hash table can become full, and if it does, it's very expensive to increase. Since the hash function divides by the table size n, increasing n changes every key's hash value. The means we must remove and reinsert every record if we resize the table.

2. Runs form as records are inserted, increasing the distance a record needs to walk from its initial hash position h during insertion.

3. Runs can merge with one another, forming very long super-runs.

Experimental analysis shows that, because of long run lengths, a table > 75% full deteriorates to $O(n)$ linear search performance. Since deletion leaves dirty locations that a search must pass over, if a table is ever > 75% full, searches will run in $O(n)$ time regardless of the number of records the table currently holds.

4.5.7 Multirecord Buckets

Another way to reduce collisions is to store more than one record in each hash table entry. For example, each entry could be implemented as an expandable array or a linked list—a bucket—capable of holding $b > 1$ records. Insertion and deletion work identically to a simple hash table, except that we no longer need to worry about exceeding the capacity of a table position.

To search for key k_t with hash value h, we load the entire bucket $A[h]$ and scan it using linear search, binary search, or whatever strategy we've implemented to try to find a target record.

Do buckets really reduce collisions? That is, for a table that can hold a fixed number of records, does reorganizing it to use buckets reduce the collision rate, compared to a simple hash table that holds one record per table entry?

If we use buckets, the packing density of A is now r/bn, where n is the table size and b is the maximum number of entries each table position can hold. Suppose we try to insert $r = 700$ records into a simple hash table with $n = 1000$ entries. Table 4.2 reports a collision rate of 28.1% for a packing density of $r/n = 700/1000 = 70\%$. Suppose we instead built a hash table with $n = 500$ entries, each of which can hold $b = 2$ records. The packing density $r/bn = 700/2 \cdot 500 = 0.7$ is the same 70%. What is its expected collision rate?

Using the Poisson equation (Eq. 4.7), we can compute the expected number of table entries that hold 0 keys, 1 key, 2 keys, and so on. Recall that Poisson uses r/n. For the simple hash table, $r/n = 700/1000 = 0.7$, and, for the hash table with buckets, $r/n = 700/500 = 1.4$.

$$\Pr(0) = \frac{0.7^0 \, e^{-0.7}}{0!} = 0.497 \qquad\qquad \Pr(0) = \frac{1.4^0 \, e^{-1.4}}{0!} = 0.247$$

$$\Pr(1) = \frac{0.7^1 \, e^{-0.7}}{1!} = 0.348 \qquad\qquad \Pr(1) = \frac{1.4^1 \, e^{-1.4}}{1!} = 0.345 \qquad (4.10)$$

$$\Pr(2) = \frac{1.4^2 \, e^{-1.4}}{2!} = 0.242$$

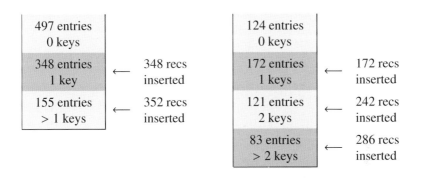

The equations and figure on the right represent the table with $n = 500$ buckets of size $b = 2$. The table has $n\Pr(0) = 124$ entries that hold no keys, $n\Pr(1) = 172$ entries that hold 1 key, and $n\Pr(2) = 121$ entries that hold 2 keys. $500 - n\Pr(0) - n\Pr(1) - n\Pr(2) = 83$ entries try to hold more than 2 keys. $700 - 172 - (2 \cdot 121) = 286$ keys hash to these positions, of which 166 are stored and $286 - 166 = 120$ collide, for a collision rate of 17.1%.

So, by simply rearranging 1000 table entries into a two-bucket table, we can reduce the collision rate from 28.1% to 17.1%, or from 197 collisions to 120 collisions.

Using multi-record buckets still poses problems for efficiency. In particular, as $r \gg n$ records are added to the table, the length of each bucket will become long, increasing the time needed for search, insertion—to check a bucket for duplicate keys—and deletion—to find the record to remove. We might be tempted to increase the size n of the hash table, but this has the same problem that we saw with progressive overflow: changing n changes the hash function, forcing us to remove and reinsert the table's records if we resize it.

CHAPTER 5

Disk-Based Sorting

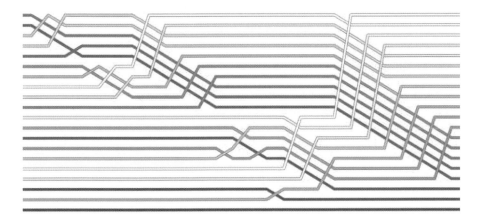

FIGURE 5.1 A weave visualization of a collection being rearranged into sorted order with mergesort

M ANY EFFICIENT algorithms exist for in-memory sorting. Can these same algorithms be used if we need to sort a file of records stored on disk? Unfortunately, two situations may preclude using an in-memory algorithm to sort a file of records.

1. The collection to sort may not fit in main memory.

2. Even if we can sort in-memory, we still need to rearrange the records in the file into sorted order, which can be very expensive.

At this point, when we talk about algorithm efficiency, we will use the number of seeks required to characterize performance. As noted previously, the cost of seeking a hard drive's disk head overwhelms other costs like rotation delay, and is much slower than most in-memory operations.

One algorithm does hold promise for disk-based sorting, however. Recall that mergesort takes collections of presorted runs, then merges them together to form

---

a fully sorted result. Mergesort starts with runs of length one, then incrementally builds longer and longer runs until the collection is sorted. It may be possible to use this strategy to read runs—or parts of runs—into main memory, reducing the number of seeks needed to read, merge, and write each run.

To make a mergesort approach work efficiently, we need to determine how to read and write runs efficiently (i.e., with as few seeks as possible).

5.1 DISK-BASED MERGESORT

To begin, assume we have a set of k presorted runs. To build a final sorted collection, we can merge the runs together.

1. Read the first part of each run from disk using an input buffer.

2. Scan the front of all k runs, pick the smallest value, remove it from its run, and write it to the sorted list.

3. Continue buffering and merging until all runs are exhausted.

Suppose each run is about the same length n/k, and that each run exhausts at about the same time. In this case, we need to do $O(nk)$ work: k comparisons to identify the smallest value, for all n values in the collection.

For small k this may be sufficient, but for large k too much time is being spent in the first step: choosing the smallest value at the front of each run. To solve this, we can store the front-of-run values in a data structure like a tournament tree. This means we can find the smallest value and update the tree in $\lg k$ time, leading to better $O(n \lg k)$ performance.

Efficient in-memory merging is important, but we also need to consider how long it takes to buffer the runs as they are ready from disk. For larger files, this is particularly important, since the cost of buffering more than a few times can quickly exceed any time spent on in-memory operations.

5.1.1 Basic Mergesort

Suppose we have 8,000,000 records, each of size 100 bytes. Further assume we have 10MB of available memory.

We can use an in-memory sort like heapsort to read subblocks of size 10MB, sort them, then write them back to disk as a presorted run. Each subblock holds 100,000 records, so we need 80 runs to process the entire file. We then perform an 80-way merge to produce a sorted file.

This approach has the following advantages.

1. The technique scales to files of virtually any size.

2. The sort/write operation is sequential. Reading from the file occurs sequentially in 10MB blocks, without the need for random access into the file. Writing the sorted runs also occurs sequentially as we append sorted values one-by-one.

3. The merge/write operation is also sequential, for similar reasons.

5.1.2 Timing

How much time is required to perform the full mergesort? We'll use the specifications for the Seagate Barracuda from Chapter 1, along with a few simplifying assumptions.

- files are stored on contiguous cylinders, and no time is needed for a cylinder-to-cylinder seek, so only one seek is needed for sequential file access, and

- data extents are staggered so only one rotation delay is required for a sequential file access.

Recall that the Seagate Barracuda has an 8 ms seek and a 4 ms rotation delay. I/O is performed during the mergesort in four steps.

1. Reading a data subblock to create a run.

2. Writing the run.

3. Reading from the sorted runs during merging.

4. Writing the sorted file.

Step 1: Read subblocks. Eighty 10MB reads are needed to create the 80 presorted runs, each requiring a seek and a rotation delay of 8 ms and 4 ms, respectively. Assuming a data transfer rate of 23300 bytes/ms, it takes $^{10}/_{0.0233} \approx 429$ ms to read 10MB, giving

$$\begin{aligned}\text{Total per run:} &\quad 8 + 4 + 429 = 441 \text{ ms}\\ \text{80 runs:} &\quad 441 \cdot 80 = 35.3 \text{ sec}\end{aligned} \tag{5.1}$$

Step 2: Write runs. The time needed to write a run is a seek, a rotation delay, and the time to transfer 10MB of data. This makes it equivalent to the time needed in Step 1 to read the data in the run: 441 ms per run, or 35.3 sec for 80 runs.

Step 3: Read and merge. To merge the runs, we divide our 10MB of main memory into 80 input buffers of size $^{10000000}/_{80} = 125000$ bytes each. It requires $^{10}/_{0.125} = 80$ sequential reads to buffer an entire 10MB run.

$$\begin{aligned}\text{Total per buffer:} &\quad 8 + 4 + 5.4 = 17.4 \text{ ms}\\ \text{Total per run:} &\quad 17.4 \cdot 80 = 1392 \text{ ms}\\ \text{80 runs:} &\quad 1392 \cdot 80 = 111.4 \text{ sec}\end{aligned} \tag{5.2}$$

Step 4: Write sorted file. Normally, we'd need to allocate part of our 10MB of memory as an output buffer. However, in this example we'll assume the OS provides two 200000 byte output buffers, allowing us to save sorted results and write results

TABLE 5.1 Increasing the size of a file by 10× increases the mergesort cost by 36.1×

	800MB	8000MB	Comparison
Step 1	35.3	352.8	800 runs of 100,000 rec/run, so 800 seeks (10× more)
Step 2	35.3	352.8	
Step 3	111.4	8025.6	800 runs @ 12,500 recs/run, so 800 seeks/run or 640,000 total seeks (100× more)
Step 4	82.4	823.4	40,000 seeks + xfer (10× more)
Total	264.4	9554.6	

× 36.1

to disk in parallel. To output the full 800MB file we need to flush the output buffer $800/_{0.2} = 4000$ times.

$$
\begin{aligned}
\text{Total per buffer:} & \quad 8 + 4 + 8.6 = 20.6\,\text{ms} \\
\text{4000 buffers:} & \quad 20.6 \cdot 4000 = 82.4\,\text{sec}
\end{aligned} \tag{5.3}
$$

The total for all four steps is $35.3 + 35.3 + 111.4 + 82.4 = 264.4\,\text{sec} = 4\,\text{min}$, 24.4 sec.

Suppose it was possible to hold the file's keys in memory and sort them. Once this was done, how much time would we need to rearrange either the index or the sorted file on disk? Assuming two seeks for each record—one to read it in the original file, and one to write it to its location in a new, sorted file—the seek and rotation delay costs alone would be $(8 + 4) \cdot 2 \cdot 8000000 = 192000\,\text{sec}$ or 53 hr, 20 min. This is $\approx 726\times$ slower than the basic mergesort.

5.1.3 Scalability

How well does the basic mergesort algorithm scale? To test this, suppose we increase the size of the file by 10×, from 800MB to 8000MB. Table 5.1 shows the cost of each mergesort step for the original and larger files. Unfortunately, the cost to sort increases by 36.1×. This means the basic mergesort is not exhibiting $O(n \lg n)$ performance.

When we examine the cost of the individual mergesort steps, it's clear that Step 3 is the problem. Further investigation shows that the read and merge part of Step 3 is where all the extra time is being spent.

In the original file, we produced $k = 80$ runs with our available memory of 10MB. When we increased the file size by 10×, we increased the total number of runs to $k' = 800$, that is, $k' = 10k$.

During the read and merge of Step 3, each input buffer is $\frac{1}{k}$ of available memory, meaning it can hold $\frac{1}{k}$ of a run. An input buffer that holds $\frac{1}{k}$ of a run will need k seeks to buffer the entire run. For the new $k' = 10k$, we've reduced the size of each input buffer by 10× to $\frac{1}{800}$ of a run. We now need $k' = 800$ seeks to buffer a run, versus the $k = 80$ seeks for the original file.

In summary, Step 1 produces k' runs that require k' seeks to read and merge in Step 3, requiring $O(k'^2)$ time, or, given $k' = 10k$, $O(k^2)$ time. k is directly proportional to n, since if we increase n by 10×, we increase k by 10× as well. Therefore, Step 3 runs in $O(n^2)$ time.

What can we do to improve the running time of Step 3? Notice that the fundamental problem is the number of runs k. Just as the $O(k^2)$ hurts us disproportionately if k is increased, it will help us if k is decreased. We will look at four strategies, some of which directly reduce k, and others that use intelligent I/O to mitigate the cost of increasing k.

1. Add more memory.

2. Add more hard drives.

3. Break the merge in Step 3 into multiple steps.

4. Increase the size of the runs produced in Step 1.

5.2 INCREASED MEMORY

Given the increase in seeks in Step 3 of the mergesort is roughly proportional to $O(k^2)$, more memory should provide a significant improvement in performance. More memory will produce fewer runs k in Step 1, which will lead to fewer seeks k over those fewer runs in Step 3.

As an example, suppose we allocated 40MB rather than 10MB in our larger example with an 8000MB file. Table 5.2 shows this increases overall performance by 4.1×, with a speedup in Step 3 of 9.7×.

5.3 MORE HARD DRIVES

Another way to directly improve performance is to increase the number of hard drives available to hold runs. This helps because it reduces the number of seeks required in each step.

For example, suppose we allocated 801 hard drives to our larger example of an 8000MB file with 10MB of available memory. In Steps 1 and 2 we would only need a total of 801 seeks: one to seek to the front of the input file stored on the first drive, and one to seek to the front of each of the 800 runs written to the remaining 800 drives. Because the drive head isn't being shared between reading data for a run and writing the run, the number of seeks is much lower.

In Steps 3 and 4 we would again only need 801 seeks: 800 to move the drive heads to the beginning of the 800 runs, each stored on a separate hard drive, and one

TABLE 5.2 Increasing the memory available during mergesort by 4× decreases cost by 4.1×

	10MB	40MB	Comparison
Step 1	352.8	345.7	200 runs of 400,000 rec/run, so 200 seeks (4× fewer)
Step 2	352.8	345.7	
Step 3	8025.6	823.4	200 runs @ 2,000 recs/run, so 200 seeks/run or 40,000 total seeks (16× fewer)
Step 4	823.4	823.4	40,000 seeks + xfer (same output buffer size)
Total	9554.6	2338.2	

$$\times 4.1$$

to seek to the front of the sorted file being written on the final drive as the runs are merged. Again, because reading and writing are performed independently, we reduce the number of seeks by ≈ 849×: from 680,000 to 801.

5.4 MULTISTEP MERGE

Because the time needed to seek overwhelms the time needed to perform an in-memory sort, we might be willing to do more in-memory sorting, if that leads to significantly fewer seeks. We'll focus on Step 3 in the mergesort. Since this step needs $O(k^2)$ seeks, reducing the number of runs k will have a significant impact on total number of seeks performed.

How can we reduce the number of runs in Step 3? One possibility is to merge subsets of runs, repeatedly merging until all the runs are processed. Consider our large example of an 8000MB file with 10MB of available memory. Steps 1 and 2 produce $k = 800$ runs. Rather than doing an 800-way merge in one step, suppose we did the following.

1. Perform 25 separate merge steps on groups of 32 runs. Each 32-way merge produces a *super-run* of length 320MB.

2. Perform a final 25-way merge on the 25 super-runs to obtain a sorted file.

This approach significantly reduces the number of seeks in the second pass, but at the expense of additional seeks and data transfer required for the first pass.[1] The question is, do the savings outweigh this additional cost?

---

[1] Notice the similarity in the idea of a multistep merge to the idea of Shell sort versus insertion sort.

Each merge in the first phase of the multistep approach combines 32 runs. We can hold $\frac{1}{32}$ of a 10MB run in its input buffer, so the entire read and merge requires $32 \cdot 32 = 1024$ seeks per merge step, or $1024 \cdot 25 = 25600$ seeks for all 25 merges.

The second phase of the merge combines the 25 super-runs, each of size 320MB. During this phase we can hold $\frac{1}{800}$ of a run in its input buffer, so the final read and merge step requires $800 \cdot 25 = 20000$ seeks.

We must also transfer every record three times rather than once (read–write–read versus one read for a one-pass merge). This requires 40000 extra seeks using the OS output buffers to write the super-runs, plus an additional 686.7 sec in transfer time.

Recall that a one-pass merge required $800 \cdot 800 = 640000$ seeks in Step 3. Our multistep merge needs $25600 + 40000 + 20000 = 85600$ seeks to read in the first phase, write the super-runs, then read in the second phase, representing a savings of $640000 - 85600 = 554400$ seeks or 6652.8 sec. Versus a time of 8025.6 sec for the one-pass merge, the multistep merge requires

$$\underbrace{8025.6}_{\text{orig cost}} + \underbrace{686.7}_{\text{extra xfer}} - \underbrace{6652.8}_{\text{seek savings}} = 2059.5 \text{ sec}$$

A simple two-phase multistep merge reduces the cost of Step 3 in the mergesort by 3.9×.

5.5   INCREASED RUN LENGTHS

Since the cost of Step 3 in the mergesort is sensitive to the total number of runs k being merged, smaller k can provide a significant improvement. One way to reduce k for a given file is to make the runs from Step 1 longer. This is what happens if we increase the amount of memory available for sorting.

Is it possible to make longer runs, even if we don't increase the available memory? In our larger example we used 10MB of memory to produce $k = 800$ runs, each of length 10MB. Suppose we could somehow create runs of length 20MB. This would reduce the number of runs to $k = 400$. Step 3 of the mergesort could then perform a 400-way merge, with each of the 400 input buffers holding $\frac{1}{800}$ of a run. The total number of seeks required would be $400 \cdot 800 = 320000$, versus the $800 \cdot 800 = 640000$ needed for 10MB runs.

In general, increasing the initial run lengths will decrease the amount of work needed in Step 3, since for any given file longer runs ⇒ fewer runs ⇒ bigger input buffers ⇒ fewer seeks.

5.5.1   Replacement Selection

To increase run lengths without increasing available memory, we will use a variation of heapsort that performs replacement selection. Rather than reading a block of records, sorting them, and writing them as a run, we will try to replace each record we output with a new record from the input file. Specifically,

1. Read a collection of records and heapsort them. This is the *primary heap*.

TABLE 5.3 Replacement selection used to generate longer runs; a 3-record heap produces an initial run of length 6

Input	Primary	Pending	Run
16 47 5 12 67 21 7 17 14 58 ...			
12 67 21 7 17 14 58 ...	5 47 16		
67 21 7 17 14 58 ...	12 47 16		5
21 7 17 14 58 ...	16 47 67		5 12
7 17 14 58 ...	21 47 67		5 12 16
17 14 58 ...	47 67	7	5 12 16 21
14 58 ...	67	7 17	5 12 16 21 47
58 ...		7 17 14	5 12 16 21 47 67
write run			5 12 16 21 47 67
58 ...	7 17 14		
...	14 17 58	7	

2. Write the record with the smallest key at the front of the heap to an output buffer.

3. Bring in a new record from the input file and compare its key with the key of the record we just wrote.

   (a) If the new key is bigger, insert the new record into its proper position in the primary heap. We can do this because the new record's "turn" in the current run hasn't come up yet.

   (b) If the new key is smaller, hold it in a *pending heap* of records. We must do this because the new record's "turn" in the current run has already passed. It will be held and written as part of the next run we construct.

4. Continue growing the current run until the primary heap is empty. At this point the secondary heap will be full, so write the current run, swap the primary and secondary heaps, and start a new run.

Table 5.3 shows an example of a 3-record heap used to process input using replacement selection. Because some of the new records read in can be moved to the current run, the result is a 6-record run, twice as large as the memory allocated to the input buffer.

Although replacement selection seems promising, two questions must be answered. First, how much longer are the runs that replacement selection produces, versus the normal read–sort–write method? Second, how much more expensive is it to generate runs with replacement selection?

5.5.2 Average Run Size

For a input buffer heap that holds n records, replacement selection will produce runs, on average, of length $2n$. This means that replacement selection generates half as many runs as an unmodified mergesort.

If the original data is partially sorted, the average run lengths will be greater than $2n$, generating even fewer runs. This case is not uncommon, and if it occurs, replacement selection is a strong candidate.

5.5.3 Cost

Recall the original cost in our large example, where mergesort created 800 runs of 100,000 records per run. Each run requires one seek for a sequential read and one seek for a sequential write, for a total of 1600 seeks, plus the cost to transfer 8000MB.

If we implement replacement selection, we clearly do not want to read and write our file and runs element by element, since this would incur a seek on each operation. Instead, we need to split available memory into an input buffer and a heap buffer. We choose a 1 : 3 split, allocating 2.5MB—enough to hold 25,000 records—to the input buffer and 7.5MB—enough to hold 75,000 records—to the heap buffer.

Given our input buffer size, it will take $8000/2.5 = 3200$ seeks to read the input file. Assuming the input buffer doubles as an output buffer for a run, and assuming it is usually full when the run ends, we will also need about 3200 seeks to write the runs. This means the first two steps of the mergesort with replacement selection need 6400 seeks, versus the 1600 needed in the original mergesort.

Since average run length is $2n$, our replacement selection runs will be about 15MB long, each holding 150,000 records. This means we generate $k = 8000/15 \approx 534$ runs total. Dividing 10MB into 534 input buffers allocates 18726 bytes per buffer. Each run needs $15000000/18726 \approx 801$ seeks to read it entirely. For $k = 534$ runs, we need a total of $534 \cdot 801 = 427734$ seeks for the 534-way merge. The original merge-sort needed 640000 seeks to perform its 800-way merge.

The overall improvement in terms of number of seeks saved is therefore

$$\underbrace{(1600 - 6400)}_{\text{run create cost}} + \underbrace{(640000 - 427734)}_{\text{merge savings}} = \underbrace{207466 \text{ seeks}}_{\text{total savings}}$$

Performing 207466 fewer seeks translates into a time saving of $\approx 1290\,\text{sec}$ or 21 min, 30 sec.

5.5.4 Dual Hard Drives

Another way to improve replacement selection is to increase the number of hard drives used from one to two.

If we do this, during run creation we can keep the input file on one drive, and store the runs on the second drive. This significantly reduces seeking; in our example we would need 535 seeks: 1 to move to the front of the input file, and 1 to move to the front of each run we write. Overlapping input and output will also provide a savings in transfer costs of up to 50%.

During the 534-way merge we swap the input and output disks. We still need seeks to read the runs into their input buffers, but writing the sorted file requires only 1 seek, versus the 4000 needed with a single drive.

The savings of 5865 seeks on run creation and 3999 seeks to write the sorted files adds up to 118 sec. Overlapping I/O during run creation could save an additional 171 sec.

CHAPTER 6

Disk-Based Searching

FIGURE 6.1 Credit card numbers are a good example of a large keyspace that can be indexed efficiently with extendible hashing

SEARCHING DATA that is too large to store in main memory introduces a number of complications. As with sorting, we focus on seeks during disk-based searching. For very large n, even a search that runs in $O(\lg n)$ seeks will be too slow. We're looking for something that runs closer to constant $O(1)$ time.

Given these needs, we specify two requirements for a disk-based search.

1. Searching must be faster than $O(\lg n)$ binary search.

2. Collections are dynamic, so insertion and deletion must be as fast as searching.

6.1 IMPROVED BINARY SEARCH

One obvious approach is to try to improve binary search to meet our requirements. For example, data is often stored in a binary search tree (BST). As noted above, however, a BST's best-case search, insertion, and deletion performance is $O(\lg n)$.

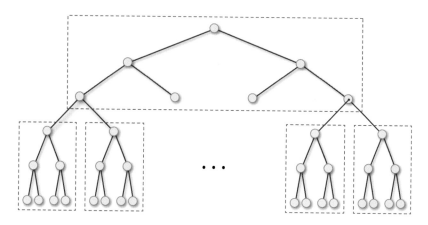

FIGURE 6.2 A paged BST with seven nodes per page and nine pages

Moreover, as data is added it is not uncommon for the BST to become unbalanced, decreasing performance toward the worst-case O(n) case.

6.1.1 Self-Correcting BSTs

Numerous improvements on BST exist. For example, AVL trees, red-black trees, and splay trees use rotations to guarantee a certain amount of balance after each insertion or deletion. Worst-case searching on an AVL tree is $1.44 \lg (n + 2)$ versus $\lg n$ for a perfectly balanced tree. A single reorganization to rebalance the tree requires no more than five parent–child reassignments, and these reorganizations are applied for approximately every other insertion, and every fourth deletion.

Unfortunately, all of this effort simply guarantees O($\lg n$) performance. Although advantageous, for disk-based search an AVL tree's height of O($\lg n$) is still "too deep" for our needs.

6.1.2 Paged BSTs

One problem with BSTs is that they require too many seeks to find a record. For disk-based searching, each seek can efficiently retrieve a sector or a cylinder of data, but most of that data is being ignored. Suppose we could reorganize a BST to store a set of locally neighboring nodes on a common disk page.

In Figure 6.2 we are storing $k = 7$ nodes per page. A fully balanced tree holds $n = 63$ nodes, and any node can be found in no more than 2 seeks. Suppose we append another level to the paged BST, creating 64 new disk pages. We can now store $n = 511$ nodes in the tree and find any node in no more than 3 seeks. Appending another level allows us to search $n = 4095$ nodes in no more than 4 seeks. Compare this to a fully balanced BST, which would need up to $\lg (4095) \approx 12$ seeks to search.

Storing $k = 7$ nodes per page was chosen to make the example simple. Real disk pages are much larger, which will further improve performance. Suppose we used disk pages of 8KB, capable of holding 511 key–reference pairs that make up a file index containing $n = 100$ million records. To find any key–reference pair, we need

$$\log_{k+1} (n + 1) = \log_{512} (100000001) = 2.95 \text{ seeks} \qquad (6.1)$$

So, with a page holding 511 nodes, we can find any record in a 100 million record index file in at most 3 seeks. This is exactly the type of search performance we're looking for.

Space Efficiency. One problem with paged BSTs is that there is still significant wasted space within each page. For example, in our $k = 7$ node example, each page has 14 references, but 6 of them point to data inside the page. This suggests we could represent the same data with only 7 keys and 8 references.

One possibility is to hold the data in a page in an array rather than a BST. We would need to perform an in-memory linear search to find records within each page, but this is relatively fast. This change would allow us to store more elements per page, potentially reducing the total number of seeks needed to locate records and therefore offering a time savings much larger than the cost of the in-memory search.

Construction. A much more significant problem with paged BSTs is guaranteeing they are relatively balanced, both when they are initially constructed, and as insertions and deletions are performed. For example, supposed we insert the following data into a paged BST that holds $k = 3$ keys in each page.

C S D T A M P I B W N G U R K E H O L J Y Q Z F X V

Figure 6.3 shows the paged BST tree that results. To try to maintain reasonable balance, we performed AVL rotations within a page during insertion to reorganize our keys. This is why the keys C, S, and D produced a page with D at the root and C and S as its left and right child nodes.

Even with page-based rotations, the tree is not balanced. The problem is that the tree is built top-down. This means the first few keys *must* go at or near the root of the tree, and they can't be moved. The root elements partition the remainder of the tree, so if those elements aren't near the center of the set of keys we're storing, they produce an unbalanced tree.

In our example, D, C, and S are stored in the root page. These keys are located near the left and right ends of the key list, so they do a poor job of partitioning the list, and therefore produce unbalanced regions in the tree. In particular, D pushes three keys to the left and 22 keys to the right. Rotating within a page can't correct this type of problem. That would require rotating keys *between* pages. Unfortunately, trying to keep individual pages balanced both internally and well placed relative to one another is a much more difficult problem.

In summary, grouping keys in individual disk pages is very good for improved search performance, but it introduces two important problems.

1. How can we ensure keys at the root of the tree are good partitioning keys for the current collection?

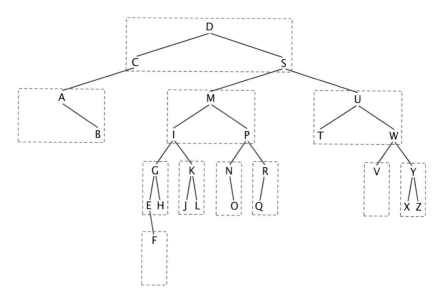

FIGURE 6.3 A paged BST with three nodes per page, constructed from letters of the alphabet in a random order

2. How can we avoid grouping keys like D, C, and S that shouldn't share a common page?

Any solution to our problem should also satisfy a third condition, to ensure each page's space is used efficiently.

3. How can we guarantee each page contains a minimum number of nodes?

6.2 B-TREE

B-trees were proposed by Rudolf Bayer and Ed McCreight of the Boeing Corporation in 1972.[1] Significant research followed in the 1970s to improve upon the initial B-tree algorithms. In the 1980s B-trees were applied to database management systems, with an emphasis on how to ensure consistent B-tree access patterns to support concurrency control and data recovery. More recently, B-trees have been applied to disk management, to support efficient I/O and file versioning. For example, ZFS is built on top of an I/O efficient B-tree implementation.

The original B-tree algorithm was designed to support efficient *access and maintenance* of an index that is too large to hold in main memory. This led to three basic goals for a B-tree.

1. Increase the tree's node size to minimize seeks, and maximize the amount of data transferred on each read.

---

[1] Organization and maintenance of large ordered indexes. Bayer and McCreight. *Acta Informatica 1*, 3, 173–189, 1972.

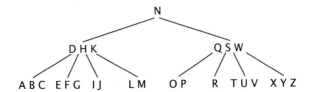

FIGURE 6.4   An order-4 B-tree with 3 keys per node, constructed from letters of the alphabet in a random order

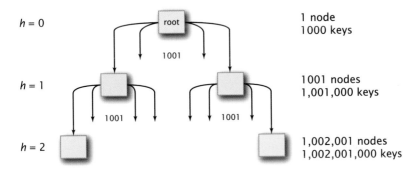

FIGURE 6.5   An order-1001 B-tree with 1000 keys per node; three levels yield enough space for about 1.1 billion keys

2. Reduce search times to a very few seeks, even for large collections.

3. Support efficient local insertion, search, and deletion.

   The key insight of B-trees is that the tree should be built bottom-up, and not top-down. We begin by inserting keys into a single leaf node. When this leaf node over-flows, we split it into two half-full leaves and promote a single key upwards to form a new root node. Critically, since we defer the promotion until the leaf overflows, we can pick the key that does the best job of partitioning the leaf. This split–promote operation continues throughout the life of the B-tree.

   A B-tree is a generalization of a BST. Rather than holding 1 key and pointers to two subtrees at each node, we hold up to $k - 1$ keys and k subtree references. This is called an order-k B-tree. Using this terminology, a BST is an order-2 B-tree. Figure 6.4 shows an order-4 B-tree used to store the same collection of keys we inserted into the paged BST in Figure 6.3.

   Although our examples have low order, a B-tree node will normally hold hun-dreds or even thousands of keys per node, with each node sized to fill one or more disk pages. Figure 6.5 shows the number of nodes and keys in the first three levels of an order-1001 B-tree. Even with a very small height of 3, the tree holds more than a billion keys. We need at most 3 seeks to find any key in the tree, producing exactly the type of size : seek ratio we are looking for.

The B-tree algorithm guarantees a number of important properties on order-k B-trees, to ensure efficient search, management, and space utilization.

1. Every tree node can hold up to $k - 1$ keys and k subtree references.

2. Every tree node except the root holds at least $\lceil k/2 - 1 \rceil$ keys.

3. All leaf nodes occur at the same depth in the tree.

4. The keys in a node are stored in ascending order.

6.2.1 Search

Searching a B-tree is similar to searching a BST; however, rather than moving recursively left or right at each node, we need to perform a k-way search to see which subtree to probe.

```
search_btree(tg, node, k)
Input: tg, target key, node, tree node to search, k, tree's order

s = 0                              // Subtree to recursively search

while s < k-1 do
    if tg == node.key[ s ] then
    |   return node.ref[ s ]       // Target found, return reference
    else if tg < node.key[ s ] then
    |   break                      // Subtree found
    else
    |   s++
    end
end

if node.subtree[ s ] ≠ null then
|   return search_btree( tg, node.subtree[ s ], k )
else
|   return -1                      // No target key in tree
end
```

The btree_search function walks through the keys in a node, looking for a target key. If it finds it, it returns a reference value attached to the key (e.g., the position of the key's full record in the data file). As soon as we determine the target key is not in the node, we recursively search the appropriate subtree. This continues until the target key is found, or the subtree to search doesn't exist. This tells us the target key is not contained in the tree.

To initiate a search, we call btree_search(tg, root, k) to start searching at the root of the tree. Searching in this way takes $O(\log_k n)$ time.

6.2.2 Insertion

Inserting into a B-tree is more complicated than inserting into a BST. Both insertion algorithms search for the proper leaf node to hold the new key. B-tree insertion must also promote a median key if the leaf overflows.

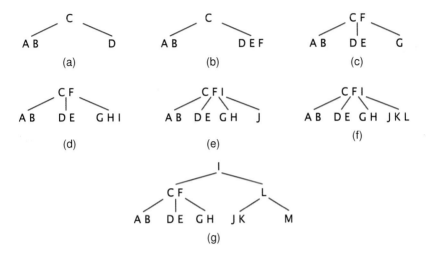

FIGURE 6.6 Inserting A B C D E F G H I J K L M into an order-4 B-tree: (a–f) insert, split and promote; (g) the final tree

1. Search to determine which leaf node will hold the new key.

2. If the leaf node has space available, insert the key in ascending order, then stop.

3. Otherwise, split the leaf node's keys into two parts, and promote the median key to the parent.

4. If the parent node is full, recursively split and promote to its parent.

5. If a promotion is made to a full root node, split and create a new root node holding only the promoted median key.

Consider inserting A B C D E F G H I J K L M into an order-4 B-tree.

- A, B, and C are inserted into the root node,

- D splits the root into two parts, promoting C to a new root node (Figure 6.6a),

- E and F are inserted into a leaf node (Figure 6.6b),

- G splits the leaf, promoting F to the root (Figure 6.6c),

- H and I are inserted into a leaf node (Figure 6.6d),

- J splits the leaf, promoting I to the root (Figure 6.6e),

- K and L are inserted into a leaf node (Figure 6.6f), and

- M splits the leaf, promoting L to the root. L splits the root, promoting I to a new root node (Figure 6.6g).

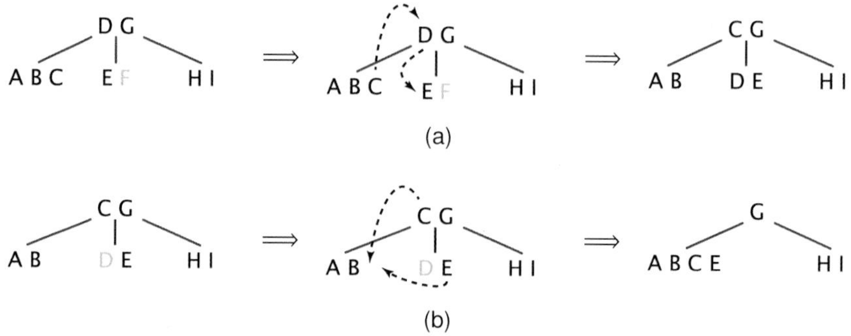

(a)

(b)

FIGURE 6.7 Deleting from an order-5 B-tree: (a) removing F produces an underfull leaf that borrows a key from its left sibling; (b) removing D produces an underfull leaf that coalesces with its left sibling and parent to form a full leaf

Notice that at each step of the insertion, the B-tree satisfies the four required properties: nodes hold up to $k - 1 = 3$ keys, internal nodes hold at least $\lceil 3/2 - 1 \rceil = 1$ key, all leaf nodes are at the same depth in the tree, and keys in a node are stored in ascending order.

Each insertion walks down the tree to find a target leaf node, then may walk back up the tree to split and promote. This requires $O(\log_k n)$ time, which is identical to search performance.

6.2.3 Deletion

Deletion in a B-tree works similarly to deletion in a BST. If we're removing from a leaf node, we can delete the key directly. If we're removing from an internal node, we delete the key and promote its predecessor or its successor—the largest key in the internal node's left subtree or the smallest key in its right subtree, respectively.

In all cases, either deleting directly from a leaf or promoting a predecessor or successor, a key is being removed from a leaf node, leaving l keys in the leaf. If $l \geq \lceil k/2 - 1 \rceil$, then we can stop. If $l < \lceil k/2 - 1 \rceil$, however, the leaf is underfull and we must rebalance the tree to correct this.

Borrow. We first check the leaf's left sibling, then its right sibling (if they exist) to see whether it has more than $\lceil k/2 - 1 \rceil$ keys. If it does, we can borrow one of the sibling's keys without making it underfull.

If we're borrowing from the left sibling, we do the following:

1. Take the key in the parent node that splits the left sibling and the leaf, the *split key*, and insert it into the leaf.

2. Replace the split key with the left sibling's largest key.

In Figure 6.7a we delete F from an order-5 B-tree. Since each node must contain at least $\lceil \frac{5}{2} - 1 \rceil = 2$ keys, this causes the leaf to underflow. The leaf's left sibling contains 3 keys, so we can borrow one of its keys, placing it in the parent and pushing the parent's split key down to the leaf. Borrowing from the right sibling is identical, except that we replace the split key with the right sibling's *smallest* key.

Coalesce. If both siblings have only $\lceil \frac{k}{2} - 1 \rceil$ keys, we coalesce the leaf with its left sibling and its parent's split key.

1. Combine the left sibling's keys, the split key, and the leaf's keys. This produces a node with $k - 1$ keys.

2. Remove the parent's split key and its now empty right subtree.

In Figure 6.7b we delete D from an order-5 B-tree. Since each node must contain at least 2 keys, this causes the leaf to underflow. Neither sibling has more than 2 keys, so we coalesce the left sibling, the parent's split key, and the leaf to form a full leaf node with 4 keys. The parent node is updated to remove its split key and the empty right subtree.

After coalescing, two checks must be made. First, if the parent node was the tree's root and if it is now empty, we make the coalesced node the new root node. Second, if the parent node is an internal node and if it now has fewer than $\lceil \frac{k}{2} - 1 \rceil$ keys, it must be recursively rebalanced using the same borrow–coalesce strategy.

6.3 B* TREE

A B* tree is a B-tree with a modified insertion algorithm.[2] B* trees improve the storage utilization of internal nodes from approximately $\frac{k}{2}$ to approximately $\frac{2k}{3}$. This can postpone increasing the height of the tree, resulting in improved search performance.

The basic idea of a B* tree is to postpone splitting a node by trying to redistribute records when it overflows to the left or right siblings. If a redistribution is not possible, two full nodes are split into three nodes that are about $\frac{2}{3}$ full. Compare this to a B-tree, which splits one full node into two nodes that are about $\frac{1}{2}$ full.

If a leaf node with $l = k$ keys overflows, and a sibling—say the left sibling—has $m < k - 1$ keys, we combine the sibling's keys, the split key, and the leaf node's keys. This produces a list of $l + m + 1$ keys that we redistribute as follows.

1. Leave the first $\lfloor \frac{l+m}{2} \rfloor$ keys in the left sibling.

2. Replace the split key with the key at position $\lfloor \frac{l+m}{2} \rfloor + 1$.

3. Store the last $\lceil \frac{l+m}{2} \rceil$ keys in the leaf.

Redistribution with the right sibling is identical, except that the first $\lfloor \frac{l+m}{2} \rfloor$ keys go in the leaf, and the last $\lceil \frac{l+m}{2} \rceil$ keys go in the right sibling.

---

[2]*The Art of Computer Programming, 2nd Edition. Volume 3, Sorting and Searching.* Knuth. Addison-Wesley, Reading, MA, 1998, 487–488.

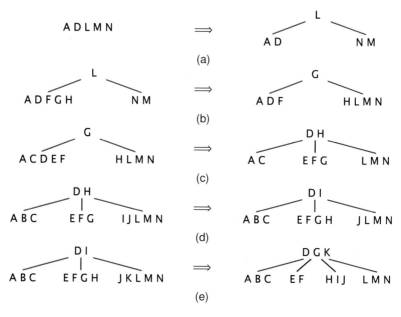

FIGURE 6.8 Inserting L A N M D F G H C E B I J K into an order-5 B* tree: (a) root node overflows and splits; (b) leaf redistributes to right sibling; (c) two full nodes split into three; (d) leaf redistributes to left sibling; (e) two full nodes split into three

If both siblings are themselves full, we choose one—say the left sibling—and combine the sibling's keys, the split key, and the leaf node's keys. This produces a list of $2k$ keys that we redistribute into three new nodes as follows.

1. Place $\lfloor {}^{2k-2}\!/_{3} \rfloor$ keys in the left node.

2. Use the next key as a split key.

3. Place the next $\lfloor {}^{2k-1}\!/_{3} \rfloor$ keys in the middle node.

4. Use the next key as a split key.

5. Place the remaining $\lfloor {}^{2k}\!/_{3} \rfloor$ keys in the right node.

If adding two split keys to the parent causes *it* to overflow, the same redistribute–split strategy is used to recursively rebalance the tree. Figure 6.8 shows an example of inserting L A N M D F G H C E B I J K into an order-5 B* tree.

- Figure 6.8a, an overfull root splits into a root and two leaf nodes,

- Figure 6.8b, an overfull leaf with $l = 5$ keys redistributes to its right sibling with $m = 2$ keys; given $l + m + 1 = 8$ keys A D F G H L N M, we place

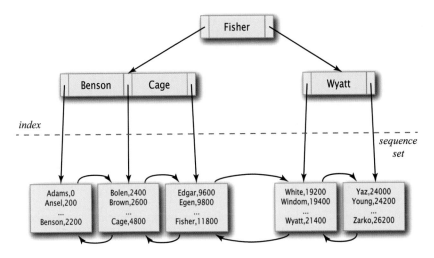

FIGURE 6.9 An order-3 B+ tree with last name acting as a key

$\lfloor {}^{l+m}\!/_2 \rfloor = 3$ keys in the leaf, the key at position $\lfloor {}^{l+m}\!/_2 \rfloor + 1 = 4$ in the parent, and $\lceil {}^{l+m}\!/_2 \rceil = 4$ keys in the right sibling,

- Figure 6.8c, an overfull leaf and its full right sibling split into three nodes; given $2k = 10$ keys A C D E F G H L M N, we place $\lfloor {}^{2k-2}\!/_3 \rfloor = 2$ keys in the left sibling, the key at position 3 in the parent, $\lfloor {}^{2k-1}\!/_3 \rfloor = 3$ keys in the leaf, the key at position 7 in the root, and $\lfloor {}^{2k}\!/_3 \rfloor = 3$ keys in the right sibling,

- Figure 6.8d, an overfull leaf redistributes to its left sibling, and

- Figure 6.8e, an overfull leaf and its full left sibling split into three nodes.

Search and Deletion. Searching a B* tree is identical to searching a B-tree. Knuth provided no specific details on how to delete from a B* tree. The problem is how to perform deletions while maintaining the constraint that internal nodes stay at least $^2/_3$ full. In the absence of a deletion strategy for B* trees, we perform B-tree deletion. This means the $^2/_3$ full requirement may be violated after deletion, but should return following a sufficient number of additional insertions.

6.4 B+ TREE

A B+ tree is a combination of a sequential, in-order set of key–value pairs, together with an index placed on top of the sequence set. The sequence set is a collection of blocks—usually disk pages—bound together as a doubly linked list. These blocks form leaf nodes for a B-tree index of keys that allows us to rapidly identify the block that holds a target key (Figure 6.9).

Placing all the data in the leaves of the index provides B+ trees a number of advantages over B-trees.

- the sequence set's blocks are linked, so scanning a range of key values requires a search for the first key and a single linear pass, rather than multiple queries on a B-tree index, and

- internal B+ tree nodes hold only keys and pointers—versus key–value pairs in a regular B-tree—so more keys can fit in each node, possibly leading to shorter trees.

The one disadvantage of B+ trees is that if the target key is found at an internal node, we must still traverse all the way to the bottom of the tree to find the leaf that holds the key's value. With a B-tree this value could be returned immediately. Since the tree's height is, by design, compact, this is normally a small penalty that we're willing to accept.

Search, insertion, and deletion into a B+ tree works identically to a B-tree, but with the understanding that all operations occur in the leaf nodes, where the key–value pairs are stored. For example, to insert a new key–value pair into a B+ tree, we would do the following.

1. Search the B+ tree to locate the block to hold the new key.

2. If space is available in the block, store the key–value pair and stop.

3. If the block is full, append a new block to the end of the file, keep the first $^k/_2$ keys in the existing block, and move the remaining $^k/_2$ keys to the new block.

4. Update the B+ tree's index (using the normal B-tree insertion algorithm) to link the new block into the index.

6.4.1 Prefix Keys

The B+ tree's internal nodes don't contain answers to search requests; they simply direct the search to a block that potentially holds a target key. This means we only need to store as much of each key as we need to properly separate the blocks.

In Figure 6.9, two keys span the gap between the first and second blocks: Benson and Bolen. The parent node uses Benson to define that the first block contains all keys ≤ Benson. We don't need the entire key to separate the blocks, however. It would be sufficient use Bf, since that would correctly define the first block to contain keys ≤ Bf, and the second block to contain keys > Bf. Similarly, the key values Cage, Fisher, and Wyatt in the internal nodes could be changed to D, G, and Y, respectively.

In general, to differentiate block A with largest key k_A and block B with smallest key k_B, it is sufficient to select any separator key k_s such that

$$k_A \le k_s < k_B \tag{6.2}$$

In our example with k_A = Benson and k_B = Bolen, k_S = Benson satisfies this requirement, but so too does the shorter k_S = Bf. If k_S has the following properties

1. k_S is a separator between k_A and k_B

2. No other separator k'_S is shorter than k_S,

then k_S satisfies the *prefix property*. B+ trees constructed with prefix keys are known as simple prefix B+ trees.[3]

By choosing the smallest k_S, we can increase the number of keys we can store in each internal node, potentially producing flatter trees with better search performance. This improvement does not come for free, however. Internal nodes must now manage variable length records, which means the number of entries in a fixed-sized node will vary. Modifications must be made to the search, insertion, and deletion algorithms to support this.

Performance. Performance results vary depending on the type of data being stored. Experimental results suggest that, for trees containing between 400 and 800 pages, simple prefix B+ trees require 20-25% fewer disk accesses than B+ trees.

A more efficient method that removes redundant prefix information along any given path in the tree produces a small improvement over simple prefix B+ trees ($\approx 2\%$), but demands 50-100% more time to construct the prefixes. This suggests that the overhead of generating more complex prefixes outweighs any savings from reduced tree height.

6.5 EXTENDIBLE HASHING

Hashing is another promising approach for searching collections that are too large to fit in main memory. In the best case, hashing can search in $O(1)$ constant time. Unfortunately, however, there are two key problems with the hash algorithms we've seen so far.

1. Search can deteriorate to $O(n)$ if too many records are inserted into the hash table.

2. There are no efficient ways to increase (or decrease) the size of a hash table.

Any hash algorithm designed to search large collections must solve both of these problems in ways that are efficient for data stored on disk.

The initial idea of extendible hashing was presented by Fagin, Nievergelt, Pippenger, and Strong of IBM Research and Universität Züirch in 1979.[4] The stated goal was to "[guarantee] no more than two page faults to locate the data associated with a given unique identifier." In other words, search requires no more than two seeks, running in $O(1)$ constant time. Equally important, the algorithm works on dynamic files, allowing data to be inserted and deleted efficiently.

---

[3] Prefix B-trees. Bayer and Unterauer. *ACM Transactions on Database Systems 2*, 1, 11–26, 1977.
[4] Extendible hashing—a fast access method for dynamic files. Fagin, Nievergelt, Pippenger, and Strong. *ACM Transactions on Database Systems 4*, 3, 315–344, 1979.

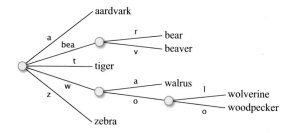

FIGURE 6.10 A radix tree containing animal names aardvark, bear, beaver, tiger, walrus, wolverine, woodpecker, and zebra

6.5.1 Trie

Extendible hashing is based on a *trie* data structure.[5,6] A trie is a tree structure that subdivides a key by its individual components. For example, if keys are words containing only lowercase alphabet letters, each level in the trie has 26 possible branches: a, b, c, and so on. A leaf in a trie contains the key that branches to the given location in the trie.

6.5.2 Radix Tree

A radix tree or Patricia trie[7] is a compressed trie where an internal node with only one child is merged with its child. This optimizes space utilization versus a standard trie.

For example, consider two keys k_0 = wolverine and k_1 = woodpecker. A trie would start with the common path w–o, then branch into two separate paths l–v–e–r–i–n–e and o–d–p–e–c–k–e–r to store the two keys.

A path in a radix tree is made only as long as is needed to differentiate each key from one another. For k_0 and k_1, a single common path wo of length 1 would exist, and it would branch into two paths l and o, also of length 1, since wo–l and wo–o are sufficient to distinguish the two keys. Figure 6.10 shows a radix tree containing wolverine, woodpecker,[8] and six other animal names.

6.6 HASH TRIES

Extendible hashing uses compressed tries to structure keys in a collection. Rather than a trie that splits letters or numbers in a key k_t, we first hash k_t to obtain h, then construct a trie that splits on the binary representation of h—a radix tree with

---

[5]Trie memory. Fredkin. *Communications of the ACM 3*, 9, 490–499, 1960.

[6]Edward Fredkin, who coined the term trie from re*trie*val, pronounced it "tree." Other authors pronounce it "try" to disambiguate it from the standard meaning of tree in computer science.

[7]PATRICIA—Practical Algorithm To Retrieve Information Coded In Alphanumeric. Morrison. *Journal of the ACM 15*, 4, 514–534, 1968.

[8]A flock of woodpeckers is called a *descent*, http://palomaraudubon.org/collective.html

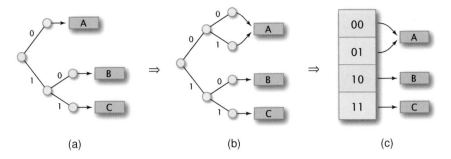

FIGURE 6.11 An extendible hash trie: (a) in its compressed form; (b) extended into a complete trie; (c) flattened into a bucket directory

a branching factor of 2. We place a fixed-size bucket at each leaf in the trie to hold keys whose hash values match the bits along the leaf's path.

Figure 6.11a shows a hash trie with three buckets. Bucket A contains records whose keys hash to 0.... Bucket B's keys hash to 10..., and bucket C's to 01....

We don't want to represent the trie using a binary tree data structure, because it will quickly grow too tall to search efficiently. Moreover, if the trie grows to exceed available memory, we will have to store it on disk using a tree-based algorithm like B-tree. If we did that, we may as well store the entire collection directly in a B-tree.

To address this need, we flatten the trie into an array representing a bucket directory. We can then index the array directly, retrieving a bucket in a single, constant-time operation. First, we must extend the trie so it forms a complete tree. This can introduce redundant information into the tree. For example, the complete trie in Figure 6.11b contains paths 00... and 01..., even though 0... is sufficient to select records in bucket A. We can see this, since internal nodes 00 and 01 both point to a common bucket.

Next, we collapse the trie into a bucket directory array A. A mimics the format of the complete trie (Figure 6.11c). Most importantly, A provides direct access to any record in the collection, based on the hash value h of its key. To search a directory A of size n for a record with key k_t,

1. Hash k_t to hash value h.

2. Extract the most significant $\lg n$ bits b of h. In our example $n = 4$, so we extract the most significant $\lg 4 = 2$ bits.

3. Convert b to an integer index i. Because of b's width, we know that $0 \le i \le n - 1$.

4. Search bucket $A[i]$ for k_t. If the record exists in A, it must be in this bucket.

For example, suppose our hash function covers the range $0 \ldots 255$ and we wanted to search the directory in Figure 6.11c for a key with hash value $h = 192$. $192 = 0b11000000$ in base 2, so we extract the most significant two bits $b = 11$,

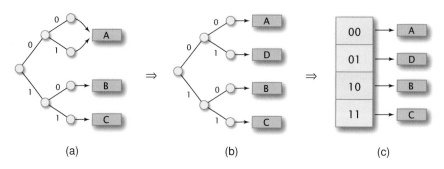

FIGURE 6.12 Inserting a new bucket into a trie: (a) insertion overflows A; (b) new bucket D inserted; (c) bucket directory

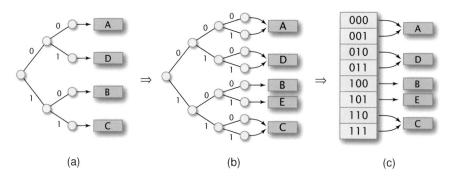

FIGURE 6.13 Extending a trie: (a) insertion overflows B; (b) trie extended to hold new bucket E; (c) bucket directory

convert them to $i = 3$, and search $A[3]$ for our record. If the key had hashed to $42 = 0b00101010$, we would have set $i = 0b00 = 0$ and searched bucket $A[0]$ instead.

6.6.1 Trie Insertion

Inserting a new record into an extendible hash table is simple. First, we use the same procedure as for searching to determine which bucket i should hold the record based on its key k_t. Next, we add the record to $A[i]$.

Remember, however, that buckets have a fixed maximum size. Keeping the maximum size small guarantees that traversing a bucket is efficient. What if there's no room in a bucket to store the new record? This represents a bucket overflow. To handle it, we must create a new bucket, link it into the proper position in the trie, and potentially move records from its sibling bucket to rebalance the trie.

When a new bucket is created, one of two things can happen. Room may exist in the trie to hold the bucket. Or, the trie itself may be full. If the trie is full, it must be extended to make room for the new bucket.

6.6.2   Bucket Insertion

As an example of bucket insertion, suppose we try to insert a new record into bucket A in Figure 6.12a. If bucket A is full, this will trigger an overflow.

In the current trie, two directory entries 00 and 01 point to bucket A. This means there is room in the trie to split A and create a new bucket D. We start by redistributing the records in A:

- records whose key hashes to 00 . . . stay in bucket A, and

- records whose key hashes to 01 . . . move to bucket D.

The new bucket D is linked into the trie by updating entry 01 to point to D. Finally, the insertion is rerun to see if there is now room to store the record.

6.6.3   Full Trie

Suppose we next try to insert a record into bucket B in Figure 6.13a. If bucket B is full, an overflow occurs. In this case, since only one trie entry references B, there is no room in the trie to split B. Instead, we must extend the trie to make room to hold a new bucket E.

To do this, we increase the number of bits used to distinguish between different hash values. Currently our trie has $n = 4$ entries, or $\lg 4 = 2$ bits per entry. We double the size of the trie to $n = 8$, extending each entry to $\lg 8 = 3$ bits.

Each pair of entries in the extended trie will point to a common bucket (Figure 6.13b). For example, entry 11 in the original trie pointed to bucket C. This means that the hash values for all records in C start with 11. In the extended trie entries 110 and 111 both point to bucket C. This is correct. Since all of C's records hash to 11 . . . , any record that hashes to 110 . . . or 111 . . . will be stored in C.

Once the trie is extended, the insertion is rerun to create a new bucket via bucket insertion, and to check if there is now room to store the record.

6.6.4   Trie Size

The ideas of inserting buckets and expanding the trie raise two important questions.

1. What stops the trie—and its corresponding bucket directory—from expanding rapidly, with numerous duplicate references?

2. Why do we hash at all? Why not use binary representations of the keys themselves to act as indices into the trie?

The answers to these questions are related. First, remember that a good hash function produces hash values that are uniformly distributed over its range. This guarantees that particular buckets won't be favored during insertion. If this is true, all buckets should fill relatively evenly, and therefore buckets should start to overflow at about the same time. This means the trie will remain relatively complete.

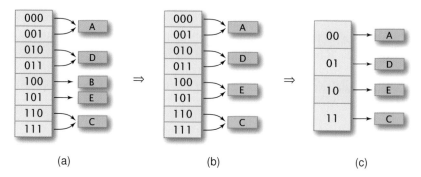

(a)   (b)   (c)

FIGURE 6.14 Deleting from a hash trie: (a) deleting from bucket B; (b) the combined records in buckets B and E underflow, collapsing to bucket E; (c) all pairs of entries point to a common bucket, allowing the trie to collapse by one bit

The need for uniform hash value distributions also explains why keys are not used to index into the trie. Most keys *do not* have a uniform distribution. If we used them directly, the trie would become unbalanced over time, making it larger than it should be to hold its records.

6.6.5 Trie Deletion

When we remove records from a hash table, it might be possible to recombine adjacent buckets. Recombining buckets may also allow us to compress the bucket directory to a smaller size.

Buckets that are candidates for recombination are sometimes called "buddy buckets." In order for buckets X and Y to be buddies,

- X must be using all available bits in the directory, that is, we can only collapse along the frontier of the trie, not in its interior,

- if X's address is $b_0 b_1 \ldots 0$, its buddy Y is at $b_0 b_1 \ldots 1$, or vice versa, and

- the combined records in X and Y fit in a single bucket.

If buckets X and Y are recombined, it may be possible to reduce the size of the bucket directory. This can happen if all pairs of directory entries $b_0 b_1 \ldots 0$ and $b_0 b_1 \ldots 1$ point to a common bucket. If this is true, then the final bit of the directory entry is not being used to differentiate between any buckets, so it can be removed and the directory's size can be halved.

Continuing with our example, suppose we delete a record from bucket B. B's directory address is 100, so its buddy is bucket E at address 101 (Figure 6.14a). If the combined number of records in B and E fit in a single bucket, B and E can be collapsed to one bucket, say bucket E (Figure 6.14b).

Now, every pair of entries in the bucket directory points to a common bucket. This means the width of the directory can be compressed from 3 bits to 2 bits. Doing this produces the 4-entry bucket directory shown in Figure 6.14c.

6.6.6 Trie Performance

Does an extendible hash trie, represented as a bucket directory array, satisfy our goal of finding key k_t with hash value h in at most two seeks? If the directory can be held in memory, we retrieve the bucket reference at $A[h]$, and use one seek to read the bucket. If the bucket directory is too large for memory, we use one seek to page in the part of the directory that holds $A[h]$ and another seek to retrieve the bucket referenced through $A[h]$.

In both cases, we move the bucket that must hold k_t from disk to main memory in no more than two seeks. This meets our goal, proving that extendible hashing has constant time search performance $O(1)$.

Another issue is the space efficiency of extendible hashing. Space is used for two data structures: the bucket directory and the buckets themselves. Analysis from the original extendible hashing paper suggests that the average space utilization is $\approx 69\%$ of the total space being allocated across all buckets. This is comparable to B-trees, which have space utilization of 67–85%. There is a periodic fluctuation in utilization over time: as buckets fill up, utilization approaches 90%. Past that point, buckets begin to split and utilization falls back to around 50%.

The directory size depends on the bucket size b and on the number of records stored in the hash table r. For a given b and r, the estimated directory size is $\frac{3.92}{b} r^{\left(1+\frac{1}{b}\right)}$.

CHAPTER 7

Storage Technology

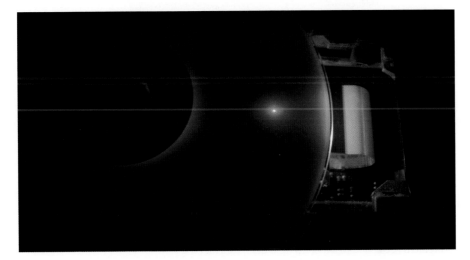

FIGURE 7.1   A DVD-RW drive operating with a 650 nm red laser[1]

S TORAGE TECHNOLOGY continues to advance, in terms of both capacity and performance. For example, Hitachi's new Ultrastar He[6] hard drives store 6TB on seven drive platters in a helium-filled enclosure.[2] Although hard drives offer an attractive storage–cost ratio, newer alternatives like optical discs and flash memory are now common. Research in areas like holographic storage, molecular memory, and magnetoresistive memory point to potential future technology that may offer significant advantages in storage capabilities.

---

[1]Felipe La Rotta, DVD burner operating with cover removed (https://commons.wikimedia.org/wiki/File:Dvd-burning-cutaway3.JPG), distributed under CC-AU-3.0 license.

[2]http://www.hgst.com/hard-drives/enterprise-hard-drives/enterprise-sas-drives/ultrastar-he6

FIGURE 7.2 An image of a Blu-ray disc surface magnified to show pits in black

7.1 OPTICAL DRIVES

Storage on optical drives like CDs, DVDs, and Blu-ray discs are based on phase change recording, which uses a laser to alter the crystalline structure of a recording media. This affects how light is reflected or absorbed when illuminated, allowing us to create a sequence of 0s or 1s on a disc's surface.

7.1.1 Compact Disc

Compact discs (CDs) were developed independently, then jointly by Philips and Sony. The first CD format and CD players were introduced in 1982. CDs are an optical system, with no read–write head. Instead, a laser source is focused and positioned via a lens onto the surface of the CD, which reflects the light back to a sensor for interpretation. Data on a CD is made up of tiny indentations, or *pits*, that follow a spiral track over the surface of the disc.

The data layer of the CD is covered by a 1.2 mm plastic substrate. This protects the disc, and since the laser is unfocused at its surface, it helps it to ignore surface dust and fingerprints. This rugged functionality is one key reason why CDs evolved into a mass market removable–transportable media.

Commercial CDs are produced using a mold of a CD's pits, and a high-speed duplication machine to stamp out copies. CD-Rs and CD-RWs allow users to write data directly. This is done by coating the surface of the disc with a phase-change material. The surface can exist in two stable states: crystalline, with the atoms arranged in a regular 3D lattice, and amorphous, with the atoms arranged in a random pattern. If the surface is heated to a few hundred degrees centigrade, the atoms rearrange into the crystalline phase. If the surface is heated to > 600°C, a melting point is achieved, and the amorphous phase results. By rapidly cooling the surface, the atoms don't have an opportunity to jump back to the crystalline state. The laser is placed in a high-power mode to produce amorphous marks, in a medium-power mode to "overwrite" a mark back to the crystalline state, and a low-power mode to read the marks.

Speed. The original CD-ROM ran at 1.2 Mb/s (125 KB/s), rotating at 200–500 rpm depending on whether the laser was positioned at the outer or inner track. This was the rate needed for audio playback. Computer-based CD-ROMs have increased rotation rates to increase the speed of data transfer. Above 12× CD-ROMs read at a constant angular velocity, so the motor does not need to change the speed of the drive as the laser is positioned to different tracks. For example, a 32× CD-ROM rotates at 4000 rpm, reading data at about 38 Mb/s on the outer track, and at about 16 Mb/s (13×) on the inner track.

Recording speeds are normally slower than reading speeds, because this depends on how quickly the recording material can change its phase. The original CD-RW drives ran at 1×, needing about 0.5 ms for phase change. Modern CD-RWs run at up to 12× recording speeds (15 Mb/s).

Audio CDs hold up to 74 minutes of sound in 333,000 sectors (2352 bytes/sector, divided into 24-byte frames), for a total capacity of 682MB. Computer-based CD-ROMs can hold up to 737MB of data on a standard CD.

7.1.2 Digital Versatile Disc

Digital Versatile Discs[3] (DVDs) followed CDs in 1995 for storing video. DVDs were invented in a collaboration between Philips, Sony, Toshiba, and Panasonic.

The original DVDs held 1.36GB of data, enough for about 135 minutes of standard definition (SD) video. The 1× transfer rate of a DVD is 11Mb/s, the rate needed to support video playback. Faster drives and multisided, multilayer discs have been introduced to increase both transfer rates and storage capacities. DVD players use a 650 nm red laser (Figure 7.1), as opposed to the 780 nm red laser used in a CD player. This produces a smaller spot, allowing smaller pits and more capacity on the disc.

Various recordable DVD formats were also introduced: DVD-RAM by Panasonic to hold up to 4.8GB of random access data, DVD-R/RW by Pioneer to hold up to 4.7GB of consumer video, and DVD+R/RW by Sony and Philips, also capable of holding up to 4.7GB of consumer video.

7.1.3 Blu-ray Disc

Digital Versatile Blu-ray discs (BD), developed to replace DVDs, are designed to hold high-definition (HD) video. Blu-ray players use a 405 nm blue laser diode. This

---

[3]DVD initially meant Digital Videodisk, but that was changed to Digital Versatile Disc to emphasize a DVD's ability to store data other than video.

was a major technical achievement, made by Sony and Philips in the late 1990s. Other important advances were also required, for example, a very hard and thin polymer used to protect the surface of consumer BDs.

Initially, two HD video recording standards existed: Blu-ray and HD-DVD, proposed by Toshiba and NEC. The key for both standards was convincing movie production studios to release movies in their format. Initially, HD-DVD was supported by Paramount, Warner Brothers, and Universal. Blu-ray was supported by Columbia, Walt Disney, and 20th Century Fox. These alliances shifted until January 2008, when Warner Brothers, the only remaining distributor of HD-DVDs, announced their movies would be released in Blu-ray. This effectively killed the HD-DVD format, as major US retail chains like Walmart dropped HD-DVD discs from their stores.

The first BD-ROM players were introduced in mid-2006. Sony's PS3 video game console, introduced in Japan and the United States in November 2006, included a Blu-ray player as standard equipment. This significantly constrained the initial production of PS3s, because of a shortage of blue laser diodes. Sony's strategy was to entrench Blu-ray as the new standard for HD video distribution by generating a large market share of players through sales of PS3s, a strategy that was eventually successful. Based on North American PS3 sales (27.3 million[4]) versus the estimated number of US households that purchased Blu-ray players (61 million[5]), approximately 45% of Blu-ray players sold are PS3s.

Standard BDs hold 25GB of data. Dual layer discs hold 50GB. A 1× BD transfers data at about 36 Mb/s. Drives currently run up to 16×, or about 576 Mb/s (72 MB/s). Recordable and rewritable variations of Blu-ray—BD-R and BD-RE, respectively–are available. Initially these formats were designed to hold video data, but they have been updated to use UDF as their file system, supporting more general data storage.

7.2 SOLID STATE DRIVES

Solid state drives (SSDs) are data storage devices that use either DRAM or non-volatile NAND flash memory to store information. SSDs are already the default storage device on smartphones, tablets, and cameras, and are now available on most laptops. Although less common, they are also used in desktop computers and storage arrays. Examples include USB keys, SDHC memory cards, and larger capacity storage arrays packaged in hard drive form factors.

DRAM-based SSDs use the same type of memory as that used in PCs and graphics cards. It offers very fast access times, on the order of 0.01 ms. Because DRAM is volatile, the drives require batteries or external power, plus a backup storage space like a traditional hard drive to copy the contents from DRAM when power is removed, then back into memory when power is restored. This is similar to hibernation sleep mode on a PC.

For flash-based SSDs, the memory is a type of EEPROM—electronically erasable-programmable read only memory. Flash memory was invented by Fujio

---

[4]http://www.vgchartz.com/analysis/platform_totals/
[5]http://hometheater.about.com/b/2013/08/06/blu-ray-disc-sales-up-15-percent-for-the-first-half-of-2013.htm

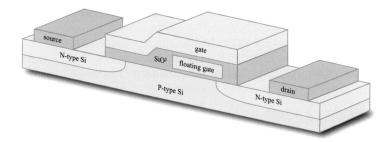

FIGURE 7.3 A block diagram of a floating gate transistor

Masuoka at Toshiba in 1980. The term "flash" was used because the block erase operation is reminiscent of a camera flash.

7.2.1 Floating Gate Transistors

Physically, flash memory is a large collection of floating gate transistors. Unlike traditional transistors, floating gate transistors are non-volatile. They can maintain their state without the need for an external charge by creating a floating gate *cage*, then using quantum tunneling to migrate electrons into or out of the cage.[6] Electrons trapped in the cage remain trapped regardless of whether power is applied, forming a binary 1 if the cell is uncharged, or a 0 if it's charged.

Floating gate transistors come in SLC—single-level cell—and MLC—multi-level cell—forms. SLC transistors have a single charge level, representing a 1—no charge—or a 0—charged. MLC transistors support multiple charge levels. For example, a 4-level MLC transistor supports four values: 11—no charge—and 10, 01, and 00—three different and distinguishable levels of charge. SLC cells are more reliable and faster, but cost more for a given amount of storage. MLC cells are cheaper, allowing increased storage density, but at the cost of potential reliability. Regardless, almost all consumer-level SSDs use MLC transistors.

Floating gate transistors have three connections: gate, source, and drain (Figure 7.3). The floating gate sits between the gate connector and the rest of the transistor. If the floating gate has no charge, a small voltage applied to the gate closes the transistor and allows it to conduct current from source to drain. If the gate contains a charge, a larger voltage is required to allow current to flow.

A NAND flash block is made up of a grid of floating gate transistors. Each row of transistors represents a *page* of memory, wired together with a *word line* that connects to the transistors' gates. The sources and drains of each column of transistors are also connected in series with a *source line* at the top and a *bit line* at the bottom. Voltages applied to the word lines, together with current applied to the transistors' source lines and measured at their bit lines allow us to read a single transistor's state. If current is

---

[6]http://arstechnica.com/information-technology/2012/06/inside-the-ssd-revolution-how-solid-state-disks-really-work

reported at the drain, then the transistor has no charge in its floating gate, representing a bit value of 1.

NAND blocks have between 32 to 256 rows (pages), and 4096 to 65536 columns, representing from 16KB to 2MB of memory. Typical page sizes are normally 8KB, which fits nicely with common OS cluster sizes of 4KB or 8KB.

7.2.2 Read–Write–Erase

NAND flash memory has a major implementation constraint: data can only be read and written one page at a time. So, to update a single byte in a page, NAND reads the page into a cache, changes the byte, then writes the page back out again.

This leads to another constraint on NAND flash memory. Data cannot be *overwritten* in place. To write data, the NAND memory must be erased to an initial state of all 1s. Individual 1s can be converted to 0s—by adding charge to the transistor's gate—but they cannot be selectively reset back to 1s. Since a high voltage is needed to erase a transistor, erase operations must be applied to an entire block.

One consequence of the overwriting and erasing constraints is that SSDs can potentially become slower over time. Initially, many empty pages exist to hold new data. Changing data can't overwrite in place, however. Instead, the old version of the page is marked as inactive, and the new version is written to a different location. As free pages decrease, it may become necessary to read an entire block into cache, erase the block, consolidate the block's active and inactive pages to make new free pages, then write it back. This read–erase–write cycle is slow, causing read operations to take more time to complete.

A third constraint on NAND memory is that it can only be written a finite number of times before it no longer functions correctly. During each block erase a very small amount of charge can become trapped in the dielectric layer that makes up the floating gate. Over time the buildup of this charge leads to a resistance that no longer allows the floating gate to change from 1 to 0. Current consumer-level SSDs allow on the order of thousands of writes.

7.2.3 SSD Controller

SSD controllers perform data striping and error correction to improve performance and guard against data corruption. They also perform a number of operations to address the issues of overwriting, block erasing, and write degradation.

First, many drives are overprovisioned to hold more memory than their stated capacity. This makes free pages available, even when the drive is nearly full.

Second, the controller is constantly performing garbage collection to keep the SSD responsive. The SSD will locate blocks with inactive pages, copy the block's active pages to new locations, then erase the entire block. This is done in the background, to try to ensure that free pages are available when needed, allowing us to avoid a costly read–erase–write cycle when data is actually being written. Modern OS's use TRIM commands to inform an SSD to mark a file's pages as inactive when the file is deleted, allowing them to be included during garbage collection.

Third, SSDs perform *wear leveling* to ensure that writes are distributed fairly evenly to all the blocks in the drive. The controller tracks how many writes each block receives, allowing it to distribute write requests throughout the drive, and to rotate static pages with active ones to ensure no blocks are underwritten.

Companies normally won't reveal exactly how their drive controllers work, since this provides them with a certain amount of competitive advantage. For example, SandForce and EMC are said to use compression and block-level deduplication to minimize the amount of incoming data that actually needs to be written to a drive. Details on how this works, and exactly how much performance improvement it provides, are not available, however.

7.2.4 Advantages

SSDs offer a number of strengths and limitations. Advantages include

- low read latency, since there is no physical seek,

- fast start-up times, since there is no spin up of a disk platter,

- high mechanical reliability, since there are no moving parts, and

- silent operation.

SSDs also have a number of disadvantages. Currently, the two main constraints are cost, which is higher than HDDs, and maximum capacity, which is lower than HDDs. For example, a 512GB SSD sells for $300 to $600, or $0.59 to $1.17/GB. A 7200 rpm 1TB HDD sells for about $70, or $0.07/GB, more than 8× cheaper per GB than the lowest cost SSD. There are also issues of asymmetric SSD read and write performance, and the potential for a limited useful lifetime.

7.3 HOLOGRAPHIC STORAGE

Holographic storage, sometimes referred to as a 3D storage method, records data as variations in a reference laser beam. Although still in a research stage, the ability to store data holographically could significantly improve both capacity and transfer rates, based on the ability to split and modulate a laser beam in parallel.

7.3.1 Holograms

To understand holographic storage, you must first understand how holograms work. A hologram is a recording of an interference pattern made during the interaction of two beams of light. This is analogous to how ripples in a pond intersect and interact. Suppose two stones are dropped in a pond. The peaks and troughs in their ripple patterns combine to amplify—when two peaks or two troughs overlap—or cancel—when a peak and a trough overlap.

Light also travels in waves, with peaks and troughs that can interact with one another in a similar manner. This technique is exploited to create a hologram. In the

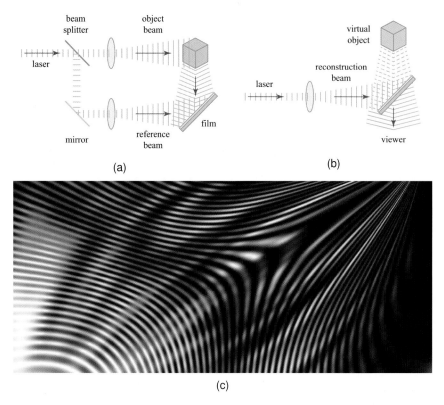

FIGURE 7.4 Holograms: (a) capturing the interference patterns between a reference beam and an object beam; (b) striking the hologram with the reference beam to extract the object beam and produce a hologram of the object; (c) a typical interference pattern

basic setup, a recording film with very high resolution is struck simultaneously by two beams of light:

- A *reference beam* shot from a laser.

- An *object beam*, a light beam from the same laser reflected off an object's surface.

The film records the interference pattern between the reference and object beams (Figure 7.4a), producing a swirl of bright ridges and dark troughs (Figure 7.4c). Laser light is used because it is stable and has coherence, both in wavelength and amplitude.

To recreate the object, the same laser light, shot in the direction of the original reference beam, is passed through the hologram (Figure 7.4b). The hologram reflects this beam in a way that "cancels out" the reference beam from the interference pattern, producing the original object beam, that is, the laser beam reflected off the object's surface. The viewer sees a reflected version of the object, floating in space. If the viewer moves his or her head, he or she sees the object from different viewpoints, since reflected light from those perspectives has also been captured in the hologram.

7.3.2 Data Holograms

In holographic storage, the same object and reference beams are used to record information as an interference pattern. Here, the object beam does not reflect off a physical object. Instead, a spatial light modulator modifies the object beam based on data values to be stored. By varying the reference beam's wavelength and angle, and the position of interference on the recording media, several thousand different holograms can be stored on a single disc.

To read data, the original reference beam reflects off the media. The reflected result is "reverse modulated" to extract the original data values that were stored. If the reference beam is diverged, the deflector can run in parallel, producing very fast transfer rates. Various manufacturers have claimed storage of up to 500 GB/in^2 and transfer of an entire file of any size up to 500 GB in 200 ms.

7.3.3 Commercialization

InPhase Technologies was the first company to try to commercialize holographic storage through their Tapestry Media system, with anticipated capacities of 500GB to 6TB. InPhase filed for bankruptcy protection in October 2011. Their assets were acquired by Akonia Holographics, which was launched as a company in August 2012.

Some of the holographic storage patents awarded to InPhase are jointly held with Nintendo of Japan. This has led to speculation that Nintendo may be the first company to introduce holographic storage for video games.

A competing technology is the Holographic Versatile Disc (HVD), initially developed in 2004. HVD combines a pair of collimated green and red lasers, using the green laser to read holographic interference patterns at the surface of a disc, and the red laser to read tracking information on an aluminum layer at the bottom of the disc. To date, no companies have chosen to implement the HVD standard.

General Electric is also working on a holographic storage system that uses Blu-ray-sized discs. A system that can record 500GB discs was demonstrated in July 2011. GE's micro-holographic storage material is said to record data at Blu-ray speeds, using technology similar to Blu-ray.

7.4 MOLECULAR MEMORY

Molecular memory is an attempt to store and retrieve data at the molecular level, with the hope of significantly increasing storage density and therefore the overall capacity of a storage device. Molecular memory is non-volatile, so it can maintain its state without power, similar to magnetic storage or flash memory.

One example of the "molecular switches" used in this memory is called rotaxane. It has a barbell shape, with a ring of atoms that can move between the two ends of

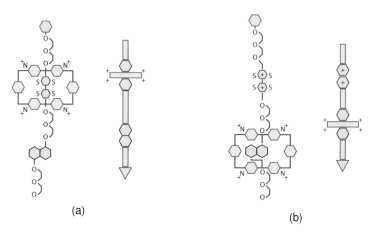

FIGURE 7.5 Bi-stable rotaxane, with the atom ring over the positive and negative endpoints of the barbell

the bar via a voltage charge (Figure 7.5). The conductivity of the molecule changes depending on where the ring is stationed, allowing two distinct states to represent binary 0 and 1.

Prototype chips sandwich a layer of molecules between two layers of nanowires. Each nanowire layer runs perpendicular to the other, forming a grid that is used to deliver voltage differences to change the state of a molecule. Rotaxane molecules are sensitive to water: one end is attracted, and the other end is repelled. This allows researchers to arrange the molecules so they are all aligned in the same direction in a 1-molecule thick layer.

Semiconductor memory has capacities of about 1.8 GB/in^2. Molecular memory has demonstrated capacities of 100 GB/in^2.

Much work remains in terms of ensuring high yields on the molecular layer, determining how to maintain the longevity of the molecules, optimizing access times, and so on. This molecular design is also being studied for semiconductors, which would allow manufacturers to replace silicon transistors with molecular transistors.

Other potential molecules are also being developed. For example, researchers at the University of Tokyo have built a molecule that changes its shape when struck by a single light beam. This design incorporates the parallel investigation of using light rather than electrical pulses to move information within a computer chip.

Work at NASA Ames has produced a *multilevel molecular memory cell* capable of manipulating three bit states rather than one. This means each cell can hold one byte. A set of molecular wires is used to chemically reduce or oxidize (redox) the molecule to set its state. Resistance is measured on a nanowire at the cell's location to read back its state. In this way a redox molecule acts like a chemical gate controlling the number of electrons it allows to pass along the wire.

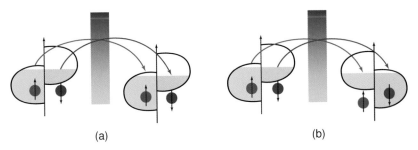

(a)   (b)

FIGURE 7.6 Spin-torque transfer: (a) when both layers have the same polarity, spin-aligned electrons pass more freely; (b) different polarity increases resistance

7.5 MRAM

Magnetoresistive random access memory (MRAM) is a non-volatile memory technology that stores information using magnetic storage elements. A memory cell is made up of two ferromagnetic plates separated by an insulating layer. One plate has a fixed magnetic polarity. The other plate's polarity can be changed. Normally, if both plates have the same polarity it represents bit 0, otherwise it represents bit 1 (Figure 7.6). Resistance at the cell can detect these states.

The most advanced MRAM uses a process known as spin-torque transfer (STT) to change the free layer's polarity. In STT, a stream of spin-aligned (polarized) electrons first passes through the fixed polarity layer, causing the electrons to spin and become polarized in the fixed layer's direction. When these spin-aligned electrons pass through the free polarity layer, they repolarize to that layer's direction. During repolarization, the electrons spin and produce a torque that sends energy through the free polarity layer. With enough energy, the free layer's polarity will reverse.

STT allows the density of cells to approach 65 nm, the minimum requirement for usable mass-market memory. MRAM is non-volatile and STT reduces the energy needed to write data. The belief is that MRAM can save up to 99% (or more) in power consumption. For reading, flash memory and MRAM have comparable power consumption. For writing, however, flash memory requires a very large voltage pulse to perform a block erase. This can also degrade the flash memory, limiting the number of write cycles it can support. MRAM has neither of these limitations.

Finally, MRAM access times have been shown to be as low as 10 ns (nanoseconds) or less. For reference, 1 ms = 1,000,000 ns, meaning we could perform 100 million memory accesses a second. Researchers suggest that, if production can be made cost efficient, MRAM could be used as a "universal" RAM, replacing all other existing types of memory storage.

CHAPTER 8

Distributed Hash Tables

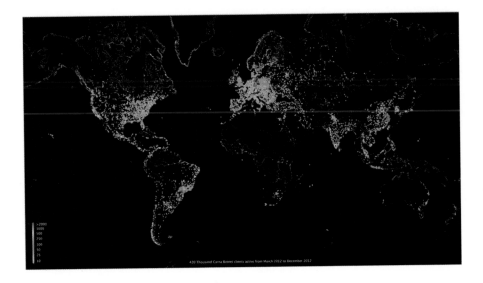

FIGURE 8.1 A visualization of Carna botnet clients in 2012;[1] botnet attacks like Storm used the Kademlia distributed hash table to organize compromised computers

D ISTRIBUTED HASH tables (DHTs) are a method of hash table lookup over a decentralized distributed file system or network. A DHT is a collection of nodes connected through some underlying topology. Key–value pairs are stored in the DHT at a parent node. Any node in the DHT can then efficiently retrieve the value by providing its key.

DHTs can act as the foundation for distributed services like P2P, DNS, and instant messaging. Notable systems that use DHTs include BitTorrent and the Storm botnet.

---

[1] Cody Hofstetter, Carna botnet March to December 2012 infection (https://commons.wikimedia.org/wiki/File:Carna_Botnet_March-December_2012.png), distributed under CC-BY-SA 3.0 license.

8.1   HISTORY

DHT research was originally motivated by P2P content distribution systems. Napster, one of the original P2P systems, maintained a centralized directory detailing each node in the network and the files it contained. This left the system vulnerable to failure, and more importantly, to litigation.

Gnutella followed Napster. It used a "flooding query" approach, where every request was multicast to every other node or machine in the network. Although robust to a single point of failure, this approach is inefficient and can generate significant wasted network bandwidth.

Freenet addressed this—as did later versions of Gnutella—by employing a localized key-based routing heuristic. In these systems, it is possible (although unlikely) that a request will fail to map a key to a value, even when that key is present in the network.

DHTs use a structured key-based routing approach to provide the decentralization of Gnutella and Freenet, and the efficiency and guaranteed results of Napster. Most modern DHTs (e.g., CAN, Chord, Pastry, and Chimera) strive to provide

- decentralization,

- scalability, and

- fault tolerance.

Much of this is achieved by requiring that a node communicate with only a few other nodes on the network. This means nodes can join and leave the network with relatively little work to update the network's state. DHTs of this type are made up of a few basic components:

- a keyspace,

- a keyspace partitioning algorithm, and

- an overlay network.

8.2   KEYSPACE

The keyspace is the description of the keys to be associated with data in the network. A common keyspace is a 160-bit string, equivalent to 20 bytes or 480 hex digits. A hash algorithm, often SHA-1,[2] is used to convert a file's key k into a hash value.

Next, a request is sent to any node in the network asking to store the file with key k. The request is propagated through the network until a node responsible for k is found. Subsequent requests to get the file with key k follow a similar pattern to locate and retrieve the file.

---

[2]The SHA-1 "secure hash algorithm," designed by the NSA and published by NIST, generates 160-bit hash values. SHA-1 has recently become vulnerable to collision attacks, and its use as a cryptographic key is being retired by many companies.

8.3 KEYSPACE PARTITIONING

Keys are assigned to a node using a keyspace partitioning algorithm. It is critical that the partitioning algorithm support adding and removing nodes efficiently.

In a basic hash table, adding or removing nodes would increase or decrease the table size, requiring all the data to be rehashed. Obviously in a dynamic distributed system this cannot happen. Instead, a form of "constant hashing" is used. Constant hashing is a technique where adding or removing one slot in the hash table does not significantly change the mapping of keys to table locations.

For example, suppose we treat keys as points on a circle. Each node is assigned an identifier i from the keyspace. Node i is then responsible for all keys "closest" to i. Suppose two neighboring nodes have IDs i_1 and i_2. The node with ID i_2 will be responsible for all keys with hash value $h \mid i_1 < h \leq i_2$.

When a node is added, it is assigned an ID in the middle of an existing node's range. Half the existing node's data is sent to the new node, and half remains on the existing node. No other nodes in the network are affected. Similarly, when a node leaves the network, it gives its data to its predecessor or its successor node.

8.4 OVERLAY NETWORK

Finally, some communication layer or overlay network is needed to allow nodes to communicate in ways that support systematically locating a node responsible for a particular key.

DHT topologies can differ, but they all share a few common properties. First, a node normally only has links to some number of local neighbors. A node does not have a view of the entire topology of the network. Second, every node will either be responsible for key k, or it will have a link to a node closer to k. This allows a greedy algorithm to simply "walk toward" k from any node until k is found.

Beyond this basic key-based routing, a few other properties are desirable. In particular, we want to enforce constraints so that the maximum number of neighbors for any node is low, and the maximum number of hops to find a node responsible for k is also low. Common values for both are $O(\lg n)$.

Robustness can be achieved by asking multiple nodes to handle the same key range. Some nodes would refer to the first node responsible for k, some to the second node, and so on.

8.5 CHORD

As a practical example, we will examine the Chord system in more detail.[3] Chord has a very simple goal. Given a key k, Chord will return a node responsible for k. It

---

[3]Chord: A scalable peer-to-peer lookup service for internet applications. Stoica, Morris, Karger, Kaashoek, and Balakrishnan. *Proceedings of the 2001 Conference on Applications, Technologies, Architectures, and Protocols for Computer Communications (SIGCOMM '01)*, San Diego, CA, pp. 149–160, 2001.

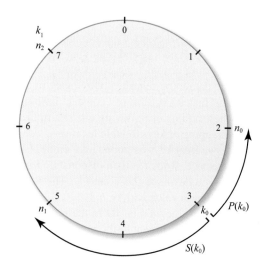

FIGURE 8.2 Chord's circular keyspace of $m = 3$ bits, generating $2^m = 8$ positions, with k_0's successor $S(k_0) = n_1$ and predecessor $P(k_0) = n_0$ highlighted

uses a DHT and associated algorithms to do this in an efficient and fault tolerant way, providing the following properties:

- **Load balancing.** Achieved by using a DHT.

- **Decentralization.** Achieved by making all nodes identical in importance within the DHT.

- **Scalability.** Achieved by ensuring key lookup is $O(\lg n)$ in the total number of nodes n.

- **Availability.** Achieved by automatically handling node joins, departs, and unexpected failures.

8.5.1 Keyspace

Chord's keyspace is a 2^m collection of m-bit keys. Both keys and nodes are assigned positions on the *keyspace circle* using the SHA-1 hash function (Figure 8.2). A key hashes directly to its position. A node hashes its IP address to get its position.

8.5.2 Keyspace Partitioning

The keyspace circle contains 2^m positions arrayed clockwise from 0 to $2^m - 1$. The successor $S(k)$ of a key k at position p_k on the circle (i.e., in the keyspace) is the first node n_i at position p_i such that $p_i \geq p_k$, that is, the first node clockwise starting from

k's position. The predecessor $P(k)$ at position p_{i-1} is the first node n_{i-1} counterclockwise starting from just before k's position with $p_{i-1} < p_k$.

For example, the keyspace in Figure 8.2 uses $m = 3$ bits to generate $2^3 = 8$ positions. Key k_0 hashes to position $p_{k_0} = 3$. The successor of k_0 is node n_1 at position $p_1 = 5$, and the predecessor of k_0 is node n_0 at position $p_0 = 2$. Key k_1 at position $p_{k_1} = 7$ has successor and predecessor $S(k_1) = n_2$ and $P(k_1) = n_1$, respectively.

Key k is assigned to the first node n_i at position p_i such that $p_i \geq p_k$, that is, k is assigned to its successor $S(k)$. In the example in Figure 8.2 k_0 is assigned to $S(k_0) = n_1$, and k_1 is assigned to $S(k_1) = n_2$.

8.5.3 Overlay Network

To support both guaranteed and efficient lookup, every node n_i maintains a local routing table with the following information:

- $P(n_i)$, and

- a finger table.

The finger table is a routing table with (up to) m entries pointing to nodes further clockwise along the keyspace circle. The j-th entry of the finger table contains the first node n_j that succeeds n_i by at least 2^{j-1} positions.

$$finger[1] = S(n_i + 2^0) \bmod 2^m$$
$$= S(n_i + 1) \bmod 2^m$$
$$= n_{i+1}$$
$$finger[2] = S(n_i + 2^1) \bmod 2^m$$
$$\ldots$$
$$finger[j] = S(n_i + 2^{j-1}) \bmod 2^m$$

Suppose node n_i is queried for the node responsible for key k. We start by checking to see if n_i or its successor is responsible for k.

- if $P(n_i) = p_{i-1} < p_k \leq p_i$, then node n_i is responsible for k (recall a key with position $p_k \mid p_{i-1} < p_k \leq p_i$ is assigned to successor $S(k)$, which is n_i), or

- if $p_i < p_k \leq p_{i+1} = finger[1]$, then give the request to $finger[1]$ (the next node counterclockwise from n_i), since we think that node is responsible for k.

Otherwise, we search the finger table for a node closer to k, on the assumption that that node will know more about the area of the keyspace circle around k. To do this, we find node n_j in the finger table whose position most immediately precedes k, then ask that node who k's successor is. n_j may know $S(k)$, or it may pass the request forward until someone answers. The finger table allows each request to jump about half as far as the previous request, so it takes $O(\lg n)$ jumps to find $S(k)$.

8.5.4 Addition

Suppose we want to add a new node n_u to the network, where n_u lies between n_t and n_v, that is, $p_t < p_u < p_v$. First, n_u obtains all keys from p_t to p_u that it is responsible for. It sets its predecessor to n_t, sets finger[1] (n_u's successor) to n_v, and inserts itself into the network.

n_t and n_v need to know that n_u has been inserted. This is not done explicitly. Instead, a stabilization protocol is run periodically. For n_v, this involves the following steps:

- ask for n_v's successor's predecessor, to see if n_v's successor (stored in finger[1]) needs to be updated, and

- inform n_v's successor about n_v's existence, allowing the successor to update its predecessor, if necessary.

When a new node n_u joins the network, it also needs to initialize its finger table, and to inform all other nodes whose finger tables are affected by n_u's addition. This is done, but it does not need to be completed before network activity resumes. Even with an out-of-date finger table, as long as successors are correct, keys will be found, just more slowly than with fully correct finger tables.[4] The same stabilization protocol is run on a node's finger table from time to time, to correct for any inconsistencies that may have arisen.

8.5.5 Failure

When node n_u fails, n_u's predecessor and finger tables pointing to n_u will need to update to use n_u's successor. To guarantee correct successor results even in the presence of failure, each node n_t maintains a successor list of the r nearest successors to n_t. When n_t notices its successor has failed, it updates its successor to the first live successor on its list. The successor list is updated during the stabilization operation.

The finger tables will slowly correct over time. Any request via a finger table for a key previously managed by the failed node will end up at n_t, and then n_t can direct the request properly to the new, correct successor.

---

[4]It's possible for a request to fail if it's issued immediately after a node joins, but before successors are stabilized. If Chord suspects this, it will reissue the request after a short delay.

CHAPTER 9

Large File Systems

FIGURE 9.1 The logo for Hadoop,[1] which was named by Doug Cutting after his son's toy elephant

T HE AMOUNT of data being captured and saved is increasing rapidly. For example, Facebook is estimated to be storing and analyzing more than 30PB (petabytes, 1 PB = 1000 TB) of user-generated data. The International Data Corporation (IDC) estimated that 1.8ZB (zettabytes, 1 ZB = 1 billion TB) of data would be created in 2011.[2] New file systems have been proposed to store and manage access to this volume of data. Issues of efficiency, cost effectiveness, and the types of analysis likely to be performed are important criteria in the design of these file systems.

9.1 RAID

RAID, originally Redundant Array of Inexpensive Disks, was proposed by Patterson, Gibson, and Katz at UC Berkeley in 1987.[3] The goal was to provide high capacity, high reliability storage from low cost, commodity PC drives. Reliability was achieved by arranging the drives into a redundant array.

---

[1]Copyright 2016 Apache Software Foundation (https://commons.wikimedia.org/wiki/File:Hadoop_logo.svg), distributed under Apache License, Version 2.0.

[2]http://siliconrepublic.com/strategy/item/22420-amount-of-data-in-2011-equa

[3]A case for redundant arrays of inexpensive disks (RAID). Patterson, Gibson, and Katz. *Proceedings of the 1988 International Conference on Management of Data (SIGMOD '88)*, Chicago, IL, pp. 109–116, 1988.

Today, RAID is Redundant Array of Independent Disks, and is a catch-all term for dividing and replicating data across multiple storage devices. Different *levels* of RAID configurations can increase reliability, increase performance, or do both.

RAID can be implemented either in hardware or in software. Hardware RAID controllers normally present the array as a single "virtual" drive, making the OS unaware of any RAID-specific operations. Software-based RAID is normally implemented in the OS. Here, applications see a single "virtual" drive, but the OS is responsible for managing the RAID array. RAID offers three basic operations.

- **Mirroring.** Data is copied to multiple drives.

- **Striping.** Data is split over multiple drives.

- **Error correction.** Redundant data is stored to detect and possibly correct data corruption.

Redundancy is achieved either by writing the same data to multiple drives—mirroring—or by calculating parity data such that the failure of one drive can be survived—error correction. Various RAID levels define which operations are performed.

- **RAID 0.** Striping, data is split to allow for parallel I/O for improved performance; no redundancy; no fault tolerance; full drive capacities are available.

- **RAID 1.** Mirroring, data is duplicated over pairs of drives, provides fault tolerance and some improved read performance; half the drive capacities are available.

- **RAID 3/4.** Striping with parity, data is split over two drives, a dedicated third drive holds parity information, provides improved performance and fault tolerance.

- **RAID 5.** Striping with parity, data is split over multiple drives, parity is interleaved with the data; usually requires 2 : 3 ratio of available space to total drive capacity; can survive failure, uninterrupted, of one drive.

The main difference between RAID 3/4 and RAID 5 is that RAID 3/4 is optimized for sequential read–write access, but not for random read–write access.

9.1.1 Parity

Parity in a RAID array is maintained with xor (exclusive or). For example, suppose we have two bytes on two drives.

$$D_1: \ 0\ 1\ 1\ 0\ 1\ 1\ 0\ 1$$
$$D_2: \ 1\ 1\ 0\ 1\ 0\ 1\ 0\ 0$$

The parity byte is the xor of the two data bytes, stored on a separate drive D_3.

$$01101101 \oplus 11010100 = 101110101 \qquad (9.1)$$

If any one drive is corrupted or fails, the missing data can be reconstructed from the other two drives. For example, if we lose data on drive D_2, we can xor D_1 and D_3 to recover it.

$$\underset{D_1}{01101101} \oplus \underset{D_3}{101110101} = \underset{D_2}{11010100} \qquad (9.2)$$

Additional RAID types exist, for example, RAID 6, which is striped with dual parity, allowing it to survive two simultaneous drive failures; RAID 1+0, a set of RAID 1 mirrored drives, themselves bundled and striped to build a mirrored RAID 0 array; or RAID-Z, a ZFS RAID array.

9.2 ZFS

The zettabyte file system (ZFS) was originally developed at Sun Microsystems by a team led by Jeff Bonwick and Matthew Ahrens in 2004.[4] A single ZFS pool can hold up to 256 zebibytes (2^{78} bytes), and a system can hold up to 2^{64} zpools, for a total theoretical capacity of about 5 sextillion ZB (2^{142} bytes). For comparison, the number of atoms in the planet Earth is estimated to be 2^{166}.

ZFS attempts to remove the need for a separate volume manager and file system. Traditionally, a volume manager is used to prepare one or more hardware storage devices to appear as a single device, and to be managed with a file system.

In ZFS, the basic storage unit is a "storage pool" or zpool. In a traditional system the volume manager would be needed to virtualize one or more physical devices into a single virtual "drive." For example, a RAID array uses this approach to map multiple physical HDDs into a single virtual storage pool. This unified view of the underlying hardware allows a file system to sit on top, without having to worry about physical or other details of the hardware itself.

In essence, ZFS replaces explicit volume management. ZFS automatically aggregates all available storage devices into a storage pool. The storage pool understands physical details about the storage hardware—devices, layout, redundancy, and so on—and presents itself to an OS as a large "data store" that can support one or more file systems. There are a number of advantages to the ZFS approach.

- file systems are not constrained to specific hardware devices,

- file systems do not need to predetermine their sizes, since they can grow by using any available space in the storage pool, and

- new storage is automatically detected and added to a storage pool, making it immediately available to any file system using the pool.

---

[4] zFS—A scalable distributed file system using object disks. Rodeh and Teperman. *Proceedings of the 20th IEEE Conference on Mass Storage Systems and Technology (MSS '03)*, San Diego, CA, pp. 207–218, 2003.

Sun engineers equated ZFS to memory in a PC. If you add more memory, it is simply "available" to any process that needs it, without any separate setup to virtualize the memory, bind it to specific processes, and so on.

9.2.1   Fault Tolerance

In a basic file system, data is overwritten in place. An unexpected failure can leave the file system in an inconsistent state, requiring utilities like fsck that attempt to identify and correct errors.

Journaling file systems offered a solution to this problem. Here, all actions are recorded in a separate journal log. If the system fails, the journal is replayed when the system restarts to try to roll the file system back to a consistent state. Data may still be lost, however, and care must be taken to either guarantee the journal is always consistent, or to fix it if it isn't.

ZFS uses indivisible transactions and copy on write to provide fault tolerance. First, data is never overwritten. It is read, modified, and rewritten to an available location. This means the system can always roll back to a file's previous state. Moreover, data is written as a single transaction that either commits (succeeds), or does not and is assumed to have failed.

Maintaining a file's previous contents as it changes also allows for automatic versioning. Since a file's old state can be made available, it is possible to revert files back to older versions by simply referencing past data previously stored in the file.

9.2.2   Self-Healing

All data in ZFS is checksummed. This means ZFS can determine when data corruption occurs. All checksumming is done within the file system, so it is completely transparent to applications storing and retrieving data.

If requested, ZFS can provide self-healing when data corruption is detected. The available types of redundancy include mirroring, and a variant of RAID 5 called RAID-Z.

9.2.3   Snapshots

At any time a read-only snapshot of the file system state can be requested. Data that has not changed since the snapshot is obtained through a reference to the snapshot. This means that the initial snapshot requires very little additional space when it is created.

As data is modified, it is copied from the snapshot, updated, and rewritten to a new, available location. In this way the file system grows relative to the number of differences between the snapshot and the file system's current state.

9.3   GFS

The Google File System (GFS) was developed by Google to provide a distributed file system with high performance, scalability, and fault tolerance.[5] GFS was designed around a number of observations about Google's workload and hardware environment.

1. **Hardware.** Component failures are the norm, not the exception, so the file system must assume and manage ongoing failures.

2. **Access.** Reading and appending to files is the norm, while overwriting data with random writes is very rare.

3. **File size.** It is inefficient to manage billions of small files; instead these should be bundled into much larger, multi-MB files.

4. **Coordination.** Coordinating between the file system and the applications can simplify the file system, without placing undue burdens on the applications.

9.3.1   Architecture

GFS is built from many inexpensive commodity components. This makes the file system cost efficient to implement, extend, and repair. GFS's design is optimized for this type of environment.

A GFS cluster includes a single *master* and multiple *chunkservers*. The master maintains file system metadata. Chunkservers are responsible for managing chunk data on their local disks. Each chunk is assigned a 64-bit chunk handle by the master when it is created. For reliability, chunks are replicated on multiple chunkservers. Applications communicate with the master for file metadata (e.g., to get a file's chunk handles), and with the chunkservers to read and write data. For example, to read a file, an application would perform the following steps.

1. Send the filename and a chunk index to the master, to obtain and cache a chunk handle and all the chunkservers—the replicas—managing the chunk.

2. Send a read request with a chunk handle and byte offset to the nearest replica.

3. Further read requests use the cached handle and replica list, removing any further communication with the master.

Chunks are sized to 64MB, which is much larger than a typical block size. This is done for several reasons, all of which revolve around the assumption of large files and large sequential reads and appends.

- reduces the need for the clients to interact with the master, since local reads and writes occur on a common chunk,

---

[5]The Google file system. Ghemawat, Gobioff, and Leung. *Proceedings of the 19th ACM Symposium on Operating System Principles (SOSP '03)*, Bolton Landing, NY, pp. 29–43, 2003.

- reduces network overhead by allowing persistent TCP connections between an application and a replica when performing operations on a common chunk, and

- reduces the amount of metadata stored on the master.

The main disadvantage of a large chunk size is the internal fragmentation that occurs for small files. Since GFS assumes these types of files are rare, this penalty is considered acceptable.

9.3.2 Master Metadata

The master maintains three types of metadata: file and chunk namespaces, the mapping of files to chunk handles, and the locations of each chunk's replicas. Namespace and file–chunk mappings are made persistent by storing them in an operation log. When a master starts (or restarts), it reads the log, then asks each chunkserver to report which chunks it is storing.

The master periodically sends a heartbeat message to each chunkserver. This allows it to verify the chunks a chunkserver is managing, to garbage collect chunks that have been freed, to update chunk replicas if a chunkserver fails, and to migrate data between chunkservers to maintain load balance.

9.3.3 Mutations

Mutations like writes or appends change either the contents or the metadata of a chunk. Every replica of a chunk needs to be mutated in a consistent way. Moreover, multiple clients may be mutating a chunk concurrently. To handle this, the master grants a *chunk lease* to one of the replicas, which is designated the *primary*. The primary is responsible for choosing an order for all mutations to a chunk. All replicas are instructed to follow this order of mutation. Leases timeout after 60 seconds, but the primary can request a lease extension during heartbeat communication with the master.

As an example, consider a client writing data to a file.

1. The client identifies the chunk it wants to write to.

2. The client asks the master for the chunk's replicas, and the identity of the primary.

3. The client pushes all data to the replicas.

4. The client sends a write request to the primary, which chooses an order of mutation and applies the mutations to its replica.

5. The primary sends the order of mutation to every other replica.

6. Once the other replicas reply they have written the data, the primary replies to the client that the operation is complete.

If errors are encountered, the client request is considered to have failed, and the modified region of the file is in an inconsistent state on some or all of the replicas. Clients are expected to retry pushing data to the replicas and requesting a write at the primary. If that continues to fail, the client will recontact the master to restart the entire write request.

The master maintains a chunk version number that increments whenever a new chunk lease is granted. The master and the chunk replicas record the new version number before any clients are given access to the chunk's primary. If a replica's chunkserver is unavailable, it will miss the new version number. When the chunkserver becomes available, its *stale chunk* will be detected by the master during heartbeat communication. The stale chunk will be garbage collected, and a new replica will be created to replace it.

9.3.4 Fault Tolerance

It is assumed that, among the hundreds of servers in a GFS cluster, some are always down. To maintain high overall system availability, GFS provides fast recovery and replication.

Both the master and the chunkservers are designed to restore their state and restart very quickly. In fact, GFS does not distinguish between normal termination and failure. Clients may need to wait briefly for a timeout on an outstanding request. If this happens, they simply reconnect to the (restarted) server, then retry the request.

Chunks are replicated across multiple chunkservers. The master is responsible for ensuring a sufficient number of replicas are available, and for creating new replicas when chunkservers fail. Each chunkserver uses checksumming to detect data corruption within a chunk on read and write requests. When errors are detected, the client is informed and asked to use a different replica. Next, the master is informed and a new replica is created. The corrupted replica is then marked for garbage collection.

The master is also replicated for reliability, by duplicating its operation log on multiple machines. If the master fails, external monitoring restarts it almost immediately. If its machine or disk fails, a new master is started on a separate machine, and the DNS tables are updated to map the master's name to the new machine's IP address.

GFS also maintains "shadow" masters to provide read-only access to data when the primary master is down. This enhances read availability for clients that are not mutating files, and that don't mind receiving data that might be slightly out-of-date.

9.4 HADOOP

Hadoop is an architecture that supports the storage, transformation, and analysis of very large datasets. Hadoop is part of the Apache software foundation (ASF), founded in 1999 to support developers release and support open-source software projects. Many companies have participated in Hadoop's code base, including Microsoft, Facebook, Google, and Yahoo!, which has contributed significant work to the core system.

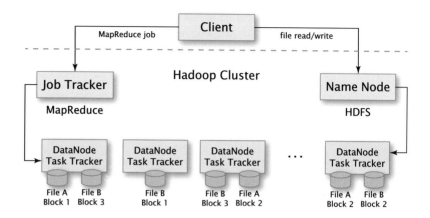

FIGURE 9.2 A typical layout for a Hadoop cluster, with a Name Node and a Job Tracker in the master, and Data Nodes managing files and MapReduce workers

Two critical components of Hadoop are the the Hadoop distributed file system,[6] (HDFS) and the MapReduce algorithm.[7] Used together, these allow efficient parallel computation on very large datasets (Figure 9.2).

9.4.1 MapReduce

MapReduce is a programming model introduced by Google to support parallelizing computation over massive datasets stored on large clusters. Not surprisingly, the model involves two steps: map and reduce.

1. **Map.** Convert input key–value pairs into intermediate key–value pairs.

2. **Reduce.** Merge intermediate key–value pairs into a result or list of results.

Consider a simple example, a very large file that stores the daily maximum temperature for 100 major cities throughout the world, for all years where such temperatures are available. Suppose we wanted to know the maximum recorded temperature for each city. One possibility would be to sort the file by city, then process it sequentially to find each city's maximum temperature. Although feasible in theory, for such a large file this would be prohibitively expensive, even with algorithms like mergesort.

Alternatively, suppose we decided to sequentially divide the file into pieces called *splits* that are small enough to fit into main memory. We distribute the splits to nodes

---

[6]The Hadoop distributed file system. Shvachko, Kuang, Radia, and Chansler. *Proceedings of the 26th Symposium on Mass Storage Systems and Technologies (MSST '10)*, Lake Tahoe, NV, pp. 1–10, 2010.

[7]MapReduce: Simplified data processing on large clusters. Dean and Ghemawat. *Proceedings of the 6th Symposium on Operating Systems Design & Implementation (OSDI '04)*, San Francisco, CA, pp. 137–150, 2004.

throughout a large cluster, assigning a map function on each node to read its split, determine which cities it contains, and identify each city's maximum temperature within the split.

```
map(name, content)
```
Input: *name*, key: split name; *content*, value: split content

```
for each city in content do
 │ max = max temp in content for city
 │ reduce( city, max )
end
```

The maximum temperatures for a given city in each split are sent to one of 100 reduce functions, also assigned to nodes within the cluster. Each reduce function is responsible for a single city's temperatures. It collects the local maximums from different map functions, then chooses an overall maximum that it returns to the calling application.

```
reduce(city, max_temp_list)
```
Input: *city*, key: city name; *max_temp_list*, value: maximum temperature in each split

```
max = -1000
for each temp in max_temp_list do
 │ if temp > max then
 │  │ max = temp
 │ end
end
return max
```

Since mapping and reducing run in parallel, we expect this algorithm to run much more quickly than a global sort and process approach.

9.4.2 MapReduce Implementation

A Communications of the Association for Computing Machinery (CACM) paper by the Google authors of the original MapReduce algorithm describe its operation in detail.[8] Assume that the map function converts a key–value pair (k_1, v_1) into an intermediate list of key–value pairs **list**(k_2, v_2) (e.g., a list of city names and associated maximum temperatures for data in a map function's split). The reduce function receives a set of intermediate key–value pairs **list**$(k_2, \textbf{list}(v_2))$ and converts them into a list of reduced values **list**(v_2), one per key k_2 (e.g., a list of maximum temperatures for each city assigned to the reduce function). To do this, the following steps are performed by the MapReduce library.

1. User-specified input files are divided into M splits that are copied to the cluster.

2. A *master* version of the user's program is initiated to monitor map functions and send intermediate results to reduce functions.

---

[8]MapReduce: Simplified data processing on large clusters. Dean and Ghemawat. *Communications of the ACM 51*, 1, 107–113, 2008.

3. The master initiates M map functions m_i on available worker nodes, then assigns responsibility for one or more splits to each m_i.

4. The master also initiates R reduce functions r_i, then assigns responsibility for one or more intermediate keys to each r_i.

5. Each map worker reads its split(s), parses the key–value pairs with the user-provided m_i, and buffers results to memory.

6. Periodically, buffered pairs are written to a map worker's local disk, partitioned into R regions based on the intermediate key(s) r_i is responsible for.

7. The buffer pair locations are returned to the master, which forwards them to the appropriate reduce workers.

8. Reduce workers use remote procedure calls to read buffered data. Once all intermediate data has been received, it is sorted by intermediate key (recall a reduce worker may be responsible for multiple keys).

9. The reduce worker walks the sorted intermediate data, passing each intermediate key and its corresponding list of intermediate values to the user-provided r_i.

10. Reduced results are appended to an output file on the cluster, producing R output files, one for each r_i.

11. Once all m_i and r_i are completed, the master returns control to the user program, allowing it to access remote worker output files as needed.

The MapReduce library includes fault tolerance, allowing it to restart failed map or reduce workers and update communication between new and existing workers. The master tries to locate map workers on the same machines that contain their splits, or on machines as close as possible to the splits, to reduce network overhead. Finally, "backup" map and reduce tasks are initiated when a MapReduce operation is close to completion, to avoid the issue of stragglers: map or reduce tasks that are slow to complete due to issues like bad hard drives, overprovisioned machines, and so on. A task can then be marked as completed when either the original or the backup worker finishes. The authors report that backup tasks can improve overall performance by 40% or more.

9.4.3 HDFS

HDFS is modelled on GFS. It stores metadata in a Name Node—similar to GFS's master—and application data on Data Nodes—similar to GFS's chunkservers. Like GFS, replication is used to provide reliability. Similar mechanisms like heartbeat communication, write leases, and journaling are used to manage mutations and support fault tolerance.

Hadoop's MapReduce algorithm is tightly integrated into HDFS. A file in HDFS is decomposed into blocks replicated throughout the cluster (Figure 9.2). When a

MapReduce job is started, splits correspond to HDFS blocks. In the optimal case, a map worker is placed on each node where a split resides, allowing local processing with little network overhead. If this is not possible, Hadoop will try to assign workers close to their splits, for example, in the same physical server rack.

Each Data Node in the cluster is restricted to a maximum number of MapReduce workers, managed with a Task Tracker (Figure 9.2). During the heartbeat communication a node informs a Job Tracker running on the master node of any available MapReduce *slots*. The Job Tracker must then schedule waiting MapReduce jobs. By default, Hadoop maintains a FIFO queue of jobs, assigning map workers as follows.

1. **Local.** If the job at the top of the queue has data local to the node, a local map worker is assigned, and any remaining slots are sent back for scheduling.

2. **Remote.** If the job at the top of the queue does not have data local to the node, a remote map worker is assigned, and scheduling stops.

The default scheduler has a number of potential drawbacks. One obvious problem is that a job at the top of the queue with unassigned map tasks blocks all other jobs, even jobs that could schedule local workers. Various suggestions have been made to improve this, for example, by temporarily delaying a job and moving it off the top of the queue when it has no local workers, or by running a matchmaking algorithm to match local workers to available MapReduce slots in some fair and optimal manner.

In its default implementation, Hadoop does not try to leverage data locality for reduce workers. A reduce worker can be assigned to any node, and data from map workers is copied to the reduce worker's node for processing. Again, various suggestions have been made to improve performance, for example, by trying to schedule map workers closer to their corresponding reduce workers, or by trying to assign reduce workers to the "center of gravity" of the map workers they will communicate with.

9.4.4 Pig

Pig[9] is a combination of Hadoop and the Pig Latin scripting language, meant to serve as a platform for analyzing large datasets. Pig and Pig Latin were developed at Yahoo! in 2007 and are used by the company (although not exclusively[10,11]) to analyze their data.

In simple terms, designers write data manipulation and analysis operations using Pig Latin. A compiler converts this code into a sequence of map–reduce operations that are run on a Hadoop server. Pig Latin is designed to support the following goals:

1. **Simplicity.** Pig Latin allows programmers to write in a high-level, procedural language that is simple to develop and maintain.

---

[9]http://pig.apache.org

[10]https://developer.yahoo.com/blogs/hadoop/pig-hive-yahoo-464.html

[11]http://yahoodevelopers.tumblr.com/post/85930551108/yahoo-betting-on-apache-hive-tez-and-yarn

2. **Optimization.** Complex data analysis tasks are automatically "parallelized" into map–reduce operations.

3. **Extensibility.** Developers can implement their own functions to perform any special-purpose processing or explicit optimization as needed.

Consider a simple example, where we want to load a file movies.csv containing movie information. In Pig, this is done with the LOAD command.

```
grunt> movies = LOAD 'movies.csv' USING PigStorage(',')
          AS (id:int, name:chararray, year:int, duration:int);
```

The PigStorage(',') option tells Pig Latin that fields are separated by commas. The list of fields following the AS command assigns names and types to each field on the rows in movies.csv. The results of the LOAD can be checked with the DUMP command, which lists the contents of the movie variable.

```
grunt> DUMP movies;
(1,The Nightmare Before Christmas,1993,4568)\
(2,The Mummy,1932,4388)\
(3,Orphans of the Storm,1921,9062)
...
(49589,Kate Plus Ei8ht,2010,)
(49590,Kate Plus Ei8ht: Season 1,2010,)
```

An important point is that Pig does not run a map–reduce job when LOAD is evaluated. Only when we issue the DUMP command is a map–reduce operation constructed and executed. Deferring map–reduce execution until you ask to expose results allows for validation and other types of checks prior to actually executing the script over a (potentially) very large dataset. For example, if we'd used the incorrect name movie_list.csv rather than movie.csv, it would be reported when we issued the DUMP command, telling us that the data file we requested didn't exist.

In addition to the standard types like int, long, float, boolean, and chararray, Pig Latin supports tuple, bag, and map. A tuple is an ordered list of fields:

```
(49589,Kate Plus Ei8ht,2010,)
```

Notice that fields can be blank, for example, Kate Plus Ei8ht has no value for the duration field in its tuple. A bag is a set of tuples:

```
{ (49589,Kate Plus Ei8ht,2010,) (2,The Mummy,1932,4388) }
```

A map is a dictionary-like [key#value] collection where chararray keys and values are joined using the # (hash) symbol.

```
[name#The Mummy, year#1935]
```

The statement above creates a map constant with two key–value pairs. The keys must always be a chararray. The values are not restricted in this way, and in our example are a chararray and an int, respectively.

Pig Latin supports numerous relational operators like JOIN (both inner and outer), FILTER, GROUP, FOREACH, and so on. This allows programmers familiar with SQL to quickly learn how to write Pig Latin scripts.

```
grunt> old_movie = FILTER movies BY year > 1900 AND year < 1960;
grunt> movie_year = GROUP old_movie BY year;
grunt> count_year = FOREACH movie_year GENERATE group,COUNT(old_movie);
grunt> DUMP count_year;
(1913,3)
(1914,20)
(1915,1)
...
(1958,73)
(1959,87)
```

DUMP is a *diagnostic* Pig Latin command, normally used for debugging a script. Other diagnostic commands including DESCRIBE, which prints the schema of a variable, ILLUSTRATE, which details how data is transformed through a series of statements, and EXPLAIN, which prints the logical, physical, and map–reduce plans Pig will use to execute a given series of statements.

```
grunt> ILLUSTRATE count_year;
```

movies	id:int	name:chararray	year:int	duration:int
		380	The General	1926	4726
		1440	Faust	1926	6425

old_movie	id:int	name:chararray	year:int	duration:int
		380	The General	1926	4726
		445	Shaft	2000	5956
		1440	Faust	1926	6425

| movie_year | group:int | old_movie:bag{:tuple(id:int,...)} |
| | | 1926 | {(380,...,4726), (1440,...,6425)} |

| count_year | group:int | :long |
| | | 1926 | 2 |

Alan Gates, an architect of Pig Latin, posted a comparison to Structural Query Language (SQL)[12] that focused on building data pipelines to retrieve raw data from

---

[12]https://developer.yahoo.com/blogs/hadoop/comparing-pig-latin-sql-constructing-data-processing-pipelines-444.html

a source, clean it, transform it, and store it in a data warehouse for later processing. He proposed five advantages of Pig Latin over SQL:

1. **Procedural.** Pig Latin is procedural, where SQL is declarative.

2. **Checkpoints.** Pig Latin allows developers to decide where to checkpoint data in the pipeline.

3. **Control.** Pig Latin allows developers to choose operator implementations rather than relying on an optimizer.

4. **Relationships.** Pig Latin supports splits in the pipeline, allowing DAG (directed acyclic graph) relationships rather than tree relationships only.

5. **Extensible.** Pig Latin allows developers to insert code in the data pipeline.

An important distinction is that Pig Latin is procedural: you describe the steps to produce what you want; where SQL is declarative: you ask for what you want and SQL determines how to provide it. The following example was provided: we want to know how many times users from different designated marketing areas (DMAs) click on a web page. This information is stored in three data sources: users, clicks, and geoinfo. In SQL, we could do something like this to return the result:

```
SELECT dma, COUNT(*)
  FROM geoinfo JOIN(
    SELECT name, ipaddr FROM users
    JOIN clicks on (users.name = clicks.user)
    WHERE click_num > 0;
  ) USING ipaddr
GROUP BY dma;
```

In Pig Latin, the result would be retrieved procedurally.

```
Users = LOAD 'users' AS (name, age, ipaddr);
Clicks = LOAD 'clicks' AS (user, url, value);
RealClicks = FILTER Clicks BY value > 0;
UserClicks = JOIN Users BY name, RealClicks BY user;
GeoInfo = LOAD 'geoinfo' AS (ipaddr, dma);
UserGeo = JOIN UserClicks BY ipaddr, GeoInfo BY ipaddr;
DMAList = GROUP UserGeo BY dma;
RealClicksPerDMA = FOREACH DMAList GENERATE group, COUNT(UserGeo);
```

In this example, the Pig Latin implementation is easier to understand for most developers, especially those who are not SQL experts. Of course, the SQL query could be decomposed into individual queries and temporary tables, but this would prevent SQL's query optimizer from executing the query in the fastest possible manner. Gates suggests that most real-world data pipelines are more complicated, and therefore will benefit more strongly from a procedural implementation.

9.4.5 Hive

Once a data warehouse is constructed, for example, by using Pig to clean, transform, and store raw data, Hive, developed at Facebook in 2008,[13, 14] can be used to assign structure to the data, and query the data using HiveQL. Similar to Pig, Hive uses HiveQL to query data on a Hadoop cluster.

HiveQL is designed to mimic the query portions of SQL, so it is a declarative language. In a Hadoop environment, the HiveQL optimizer converts a query into map–reduce jobs. This allows developers who are familiar with SQL to quickly generate HiveQL code that leverages the power of map–reduce on a data warehouse stored on a Hadoop cluster.

Hive data is stored in structured tables, similar to a relational database management system (RDBMS). This means that, like an RDBMS, an explicit *schema* describing a table's columns must be provided. Each table can be *partitioned* or sliced based on one or more of its columns.

```
CREATE TABLE movie {
    id INT, name STRING, duration INT
}
PARTITION BY {
  year INT
};
```

Notice that `year` (and any other fields in the `PARTITION BY` block) are considered columns in the table. The purpose of partitioning is to efficiently distribute data over the HDFS cluster. Specifically, each partition value (or value tuple) creates a separate subdirectory to hold records with the given value. When queries filter the data, for example, with a `WHERE` clause, only those subdirectories that match the filter need to be processed. This can significantly improve performance.

Alternatively, Hive can subdivide data into *buckets*. This technique is also meant to optimize performance, by hashing values for a target column, then distributing those hash values over a user-chosen number of buckets.

```
CREATE TABLE movie {
    id INT, name STRING, duration INT
}
PARTITION BY {
  year INT
}
CLUSTERED BY (name) INTO 20 BUCKETS;
```

Here, the movie's name is converted to a hash value, presumably in the range $0 \ldots 19$. The resulting 20 buckets are stored over the HDFS cluster in a way that makes it efficient to retrieve the contents of an individual bucket. Now, if a user

---

[13]http://hive.apache.org
[14]https://www.facebook.com/notes/facebook-engineering/hive-a-petabyte-scale-data-warehouse-using-hadoop/89508453919

queries a movie by its name, the name will be hashed and matched to a specific bucket. Only this target bucket needs to be searched.

Hive is targeted at situations where structured tables and SQL-like queries are most effective. Examples include connecting to business intelligence tools, or allowing data analysts to run ad-hoc queries on the data tables. Another distinction that is often proposed is that Pig is meant for ETL (extract–transform–load) pipelines, producing tables that Hive can then use for report generation and exploratory queries.

In spite of their differences, at their core Pig and Hive are both designed to reduce the complexity of writing map–reduce jobs, which would normally be done in a language like Java. Java map–reduce jobs will usually execute faster, but Pig and Hive programs will be usually be quicker and easier to implement.

9.5 CASSANDRA

Cassandra is a distributed database system designed to manage large structured, semi-structured, or unstructured data collections. Originally implemented in 2008 at Facebook by Avinash Lakshman and Prashant Malik,[15] Cassandra was migrated to the open source Apache Cassandra project, currently at version 3.4 as of March 2016.

Cassandra was initially built to solve Facebook's Inbox Search problem, offering Facebook users the ability to query their Facebook Inbox in two different ways: term search, where a user searches for messages with a target term, and interaction search, where a user searches for messages to and from a particular Facebook friend.

Cassandra assumes a number of design approaches similar to other systems. For example, Cassandra is built to run on clusters of commodity hardware where a subset of the machines is expected to be down at any given time. This is similar to GFS's design, and, like GFS, Cassandra uses replication to address this problem. Cassandra also avoids a centralized master node, instead arranging its nodes in a ring-like structure. Each node is responsible for a range of hashed key values, and every node understands the ring's structure so it can communicate directly with other nodes responsible for a given key. This allows a client to send a read or write request to any node in the cluster. The decentralized ring approach is similar to how distributed hash tables implement P2P communication, for example, in the Chord DHT.

Cassandra's original implementation was built to ensure performance, reliability, efficiency, and perhaps most importantly, scalability. A Cassandra data store is made up of a table, partitioned into rows. Each row includes a unique row key, together with one or more column values. Columns can be combined into sets called *column families*. A simple column family contains one or more raw columns. A super column family contains within itself one or more additional column families, supporting hierarchical, level-of-detail organization of the raw columns.[16] As an example, to support an interaction Inbox Search for conversations between a user with Facebook ID u and other users $u_1, \ldots, u_n$, u represents the user's row key, and each

---

[15]Cassandra—A decentralized structured storage system. Lakshman and Malik. *ACM SIGOPS Operating Systems Review 44*, 2, 35–40, 2010.

[16]Apache Cassandra has removed super columns, and renamed column families as *tables* (see Table 9.1 for a comparison of old and new Cassandra terminology).

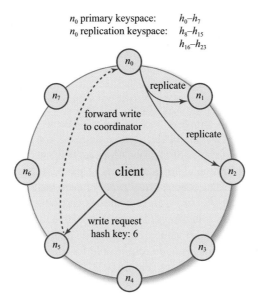

n_0 primary keyspace: h_0-h_7
n_0 replication keyspace: h_8-h_{15}
$h_{16}-h_{23}$

FIGURE 9.3 A Cassandra ring cluster, with a client's write request sent to the nearest node, forwarded to the coordinator node responsible for the new row's hashed key value, and replicated on two replication nodes

$u_i, 1 \le i \le n$ represents a simple column family. Within the simple column family u_i, raw columns contain message IDs, one for each message shared between users u and u_i. In this way, a row encapsulates all conversations between a user u and all other users in u's Inbox.

Interestingly, different rows do not need to store values for the same columns or column families. Two rows may contain both common and unique columns. This is apparent in the interaction Inbox Search example: two different users u_j and u_k most likely talk to different friends, so the simple columns in their rows will not contain the same set of user IDs. This makes Cassandra a type of NoSQL database, since it is not constrained by a fixed schema that requires each row to maintain a common column structure. This is similar to the functionality provided by MongoDB or Apache CouchDB.

9.5.1 Design

Cassandra was implemented in Java, using a number of external libraries to support fault detection, communication between nodes on the ring, blocking and non-blocking I/O, and identification of leader nodes to manage data replication.

As noted above, a Cassandra table is made up of one or more rows, each with its own primary row key. Row keys are hashed with a consistent, order-preserving hash function. The primary row key's hash value h_i identifies the node used to store the row's values. Nodes are arranged in a ring, with successive nodes n_j responsible for ranges of hash values $n_{j,\text{lo}}, \ldots, n_{j,\text{hi}}, n_{j,\text{lo}} = n_{j-1,\text{hi}} + 1$. Consistent hashing is used to ensure the removal or addition of a node only affects a local neighborhood of node values. To maintain load balancing across nodes, lightly loaded nodes are identified and moved on the ring (i.e., have their hash ranges changed and rows redistributed as needed) to support heavily loaded nodes.

Replication. To maintain high availability, Cassandra replicates each row using a replication factor f. In its simplest "rack unaware" scheme, the initial node responsible for a row is tagged as the row's *coordinator*. The $f - 1$ successor nodes of the coordinator are assigned copies of the row. More sophisticated replication strategies like "rack aware" (replication within a server rack) and "data center aware" (replication within a geographic data center) use a *leader* node to assign replication ranges to other nodes in the cluster.[17] Every node is aware of every other node, in particular the ranges of primary and replicated keys they are responsible for. This allows a client to request a row with a given hash value from any node in the cluster. The request will be forwarded to the coordinator node responsible for the key, and to all secondary nodes that replicate the key's values (Figure 9.3). Various approaches (most recent write, quorum, and so on) can be used to decide how and when to return row values to the client.

Figure 9.3 shows an example of a client requesting to write a row whose primary row key hashes to $h = 6$. Assuming a hash range of $0 \ldots 63$ and eight nodes in the cluster, each node is initially responsible for eight hash keys: $0 \ldots 7$, $8 \ldots 15$, $\ldots$, $55 \ldots 63$. In our example the replication factor is $f = 3$, with a "rack unaware" assignment for replicating nodes. The client passes its write request to the nearest node, which forwards it to the coordinator node n_0 responsible for hash value 6. The coordinator replicates the row on its replication nodes n_1 and n_2 (i.e., the two successor nodes of n_0).

Failure Detection. Cassandra uses a version of accrual failure detection[18] to identify nodes that may have failed. A value Φ is defined to identify the probability (or *suspicion*, as Cassandra calls it) that a node has failed. Once a node is assumed to have failed, it is removed from the list of active nodes. Replicas are then responsible for managing rows assigned to the failed node. If and when the node returns with a heartbeat message, it is reintegrated into the ring.

When a node is assumed to have failed, any updates to that node are queued and issued to the node if it returns. In this way Cassandra implements *eventual consistency*, or specifically a particular type of eventual consistency known as *monotonic write consistency*: write requests by a given process are guaranteed to be serialized, and if no new updates are made to a row, eventually all nodes will return a consistent set of values for that row.

---

[17] Apache Cassandra has changed replication strategies: the default strategy is now SimpleStrategy; the rack aware strategy has been depreciated as OldNetworkTopologyStrategy, and a new *NetworkTopologyStrategy* has been added to better support multiple data centers.

[18] The ϕ accrual failure detector. Défago, Urbán, Hayashibara, and Katayama. In *RR IS-RR-2004-10, Japan Advanced Institute of Science and Technology*, pp. 66–78, 2004.

Node Addition. When a new node is added to the ring, it chooses its position in various ways: randomly, based on load balancing, and so on. The new node then splits the key range of an existing successor or predecessor node. A bootstrapping process is used to copy data from the existing node to the new node, to choose f new replicas and copy data from existing replicas to new replicas, and finally to update the key ranges for the existing and new replicas within the ring.

Persistence. A node's data is stored on its local file system to maintain persistence. Write operations first write to a commit log, then update an in-memory *memtable* data structure representing rows most recently written to the node. Based on size and the number of rows being managed, the in-memory data structure is periodically dumped to disk as an *SSTable*. Writes are performed sequentially, and update an index that is used for efficient row lookup based on a row's key. The indices are persisted along with the row data in the SSTable files. As the number of SSTables grows, a background compaction algorithm is run to combine multiple SSTables into a single SSTable file.

Read operations first query the in-memory data structure, and if the target row is not found, move on to the SSTables to locate the row. Files are examined in order of newest to oldest. Filters are maintained to ensure file searches only occur when a key is located in a file. Indexes are used to guarantee column and column family values can be retrieved efficiently.

Performance. At the time the original Cassandra paper was written, the Facebook authors reported that it was used to support 250 million users running on a 150-node cluster storing 50TB of data. Both interaction and term searches ran, on average, in 10s of milliseconds, with a maximum latency of 44.41 ms for term searches.

9.5.2 Improvements

The Facebook codeblock was migrated to an Apache Cassandra open source project (hereafter Cassandra refers to Apache Cassandra) in 2010. As of June 2016, it has been steadily updated to version 3.6.[19] One important change is in the terminology used for the different components within Cassandra. Table 9.1 details the old and new terminology now being employed to describe Cassandra's components.

Numerous improvements have been added to Cassandra since it became an open source project. Cassandra has provided a magnitude of improvement in performance over the original Facebook version. The ability to delete data was added. The Cassandra Query Language (CQL) was developed to allow command-line and GUI management of a Cassandra cluster. Virtual nodes are now used in place of ordered hashing for load balancing. A new NetworkTopologyStrategy replication method is available for replicating across clusters deployed at different physical data centers. This allows for the loss of an entire data center while still maintaining data availability. Finally, the performance of various compaction algorithms has been improved.

---

[19]http://cassandra.apache.com

TABLE 9.1 Original Facebook versus current Apache terminology for Cassandra components

Facebook	Apache
row	partition
column	cell
cell name, component, value	column
cells with a shared primary key	row
simple column family	table
super column family	removed from Cassandra
rack unaware replication	`SimpleStrategy` replication
rack aware replication	`OldNetworkTopologyStrategy` replication
datacenter aware replication	`NetworkTopologyStrategy` replication

9.5.3 Query Language

Cassandra initially provided a *Thrift API* to support database management. One of the important additions when Cassandra became an Apache project was the introduction of the Cassandra Query Language (CQL). Cassandra does not implement the full SQL specification. Instead, it provides a more limited set of SQL-like commands to allow for creation and manipulation of data in Cassandra tables.

As an example of using CQL to build a Cassandra table with *wide rows* (rows that allow for different types and/or numbers of columns; Cassandra now claims it can hold up to two billion columns in a single row[20]), consider the following simple example. We first create a namespace to hold tables, then create a data table to store sensor data.

```
cql> CREATE KEYSPACE sensor
        WITH REPLICATION = {
        'class' : 'SimpleStrategy', 'replication_factor' : 3
        };
cql> USE sensor;
cql> CREATE TABLE data {
        id int, date timestamp, volts float,
        PRIMARY KEY( id, date )
     };
cql> INSERT INTO data (id,date,volts) VALUES (1,2013-06-05,3.1);
cql> INSERT INTO data (id,date,volts) VALUES (1,2013-06-07,4.3);
```

---

[20]https://wiki.apache.org/cassandra/CassandraLimitations

```
cql> INSERT INTO data (id,date,volts) VALUES (1,2013-06-08,5.7);
cql> INSERT INTO data (id,date,volts) VALUES (2,2013-06-05,3.2);
cql> INSERT INTO data (id,date,volts) VALUES (3,2013-06-05,3.3);
cql> INSERT INTO data (id,date,volts) VALUES (3,2013-06-06,4.3);
cql> SELECT * FROM data;
 key |    date    | value
-----+------------+-------
   1 | 2013-06-05 |   3.1
   1 | 2013-06-06 |   4.3
   1 | 2013-06-07 |   5.7
   2 | 2013-06-05 |   3.2
   3 | 2013-06-05 |   3.3
   3 | 2013-06-06 |   4.3
```

Notice that we defined a *composite* primary key (id, date) when we created the data table. The first part of the primary key id acts as a *partition key* (or row key in the original Cassandra terminology). The second part of the primary key date acts as a *clustering key*. This allows a user, for example, to search for all readings from a particular sensor using the partition key id, or for readings from a particular sensor on a particular day using both the partition and clustering keys. Primary and clustering keys can also affect how data is distributed throughout a ring. For example, given our explanation of how Cassandra uses a hashed primary key value to identify a partition's coordinator node, we can see that all partitions (rows) with the same primary key (row key) will be stored on a common node.

Cassandra is now used by a wide range of companies, including Apple, Facebook, IBM, Netflix, and Reddit, among others. According to Wikipedia, Cassandra became the ninth most popular database in September 2014.[21]

9.6 PRESTO

Presto[22] is query language designed to support rapid response on large external databases like those stored in a Hadoop cluster. Originally designed by Facebook, Presto supports real-time queries on petabyte-scale datasets. Unlike a language like Hive, which is best suited to large-scale, batch-based map–reduce jobs, Presto strives for interactive queries with low latency. Presto is implemented to support a subset of SQL as its query language, due to SQL's familiarity for many users.

To achieve its goal of interactive, ad-hoc query support, Presto's architecture is optimized for responding rapidly to queries in standard SQL (Figure 9.4). Based on Facebook's description in their blog post,[23] a Presto query passes through the following steps:

---

[21] https://en.wikipedia.org/wiki/Apache_Cassandra.

[22] http://presodb.io/

[23] https://www.facebook.com/notes/facebook-engineering/presto-interacting-with-petabytes-of-data-at-facebook/10151786197628920/

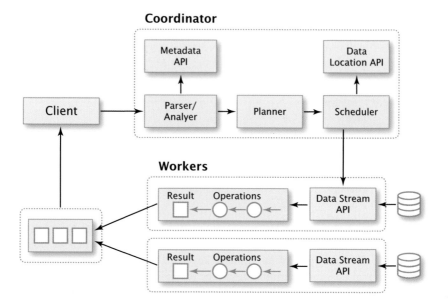

FIGURE 9.4 Presto's architecture: a client originating a query; a coordinator to parse, analyze, and schedule the query as a set of tasks; workers for each task; and results returned to the client

1. A client issues an SQL query.

2. The SQL query is sent to the Presto coordinator.

3. The coordinator parses and analyzes the query to define a plan for its execution.

4. The Presto scheduler defines a (usually parallel) execution pipeline, assigns different tasks to worker nodes closest to the data associated with each task, and monitors task progress.

5. The client requests data from an output source, which triggers results to be pulled from the different external stages where they have been collected.

   Presto's approach to executing a query is significantly different from that of a system like Hive. HiveQL converts a query into a sequence of map–reduce jobs. Although each map–reduce step is run in parallel, the sequence of jobs is run in serial, with data being read from disk at the beginning of a job, and results being written to disk at the end. This represents a significant I/O overhead, even on a Hadoop cluster optimized for efficient data management. Presto does not use map–reduce. Instead, its coordinator customizes a schedule of tasks for a particular query, with groups of tasks being parallelized when possible, and results flowing from one task to another across the network. This approach was designed to minimize disk-based I/O. The result is a significant reduction in latency for many types of queries.

Presto is also built to be extensible for different types of large file system backends. For example, Facebook uses Presto to query both HBase distributed databases and their proprietary Facebook News Feed. To support different storage architectures, Presto provides an abstraction layer that allows storage plugins (or *connectors*, in Presto's terminology) that provide a specific interface for fetching metadata, identifying data locations, and accessing raw data as needed. This is sufficient for Presto's parser, analyzer, and scheduler to construct and execute query plans. Presto currently includes numerous connectors, including support for Hive, MongoDB, MySQL and PostgreSQL, and Cassandra.[24]

Connectors are used to define *catalogs* using Presto's configuration files. The specific content of the file depends on the type of connector being created. For example, a MySQL connector defines a unique catalog name for the target server, a server URL, and the userID and password needed to access databases (called SCHEMAS in Presto) on the server. A simple MySQL configuration file (mysql.properties) for the catalog mysql would look like the following.

```
connector.name=mysql
connector-url=jdbc:mysql://server.url:3306
connection-user=root
connection-password=root-password
```

A similar approach is used for other back-end architectures. For example, to connect to a Hive data warehouse hive running on a Hadoop server, a configuration file hive.properties with a unique catalog name and URL for the Hive metastore Thrift service could be created.

```
connector.name=hive
hive.metastore.uri=thrift://server.url:9083
```

Once catalogs are defined, they can be accessed, examined, and queried using standard SQL. For example, to look at the schemas (databases), tables, and table formats for the mysql catalog, the following commands could be used.

```
presto> SHOW SCHEMAS FROM mysql;
Schema
------
local
remote
web
(3 rows)

Query 20150818_064410_00003_837eu, FINISHED, 1 node
Splits: 2 total, 2 done (100%)
0.00 [3 rows, 61B][25 rows/s, 524B/s]
```

---

[24]https://prestodb.io/docs/current/connector.html

```
presto> SHOW TABLES FROM mysql.web;
Table
------
clicks
(1 row)

Query 20150818_064532_00004_837eu, FINISHED, 1 node
Splits: 2 total, 2 done (100.00%)
0:00 [1 rows, 28B] [32 rows/s, 613B/s]

presto> DESCRIBE mysql.web.clicks;
| Column      | Type       | Null | Key | Default |
+-------------+------------+------+-----+---------+
| Language    | char(6)    | NO   |     | en      |
| Title       | char(256)  | NO   | PRI |         |
| Avg_Hits    | float      | NO   |     | 0       |
| URL         | char(256)  | NO   |     |         |
(4 rows)

Query 20150818_064691_00005_837eu, FINISHED, 9 nodes
Splits: 177 total, 177 done (100.00%)
0:12 [4 rows, 68B] [21 rows/s, 601B/s]
```

Facebook released Presto company-wide in the spring of 2013, running on clusters of up to 1,000 nodes, and processing 30,000 daily queries across a petabyte of data at multiple geographic locations. Results have been positive. Facebook's blog post notes that Presto is often up to 10× faster than Hive in terms of both CPU efficiency and latency.

Since Facebook has released Presto as an open source project, various companies have adopted it within their data analytics workgroups, including Netflix,[25] Airbnb, and Dropbox. These companies are now contributing improvements and updates to the project.

Presto is not the only interactive query language being developed for large data storage clusters. One example is Apache Spark,[26] which is designed to support operations like interactive query, streaming, machine learning, and batch processing. Another is Cloudera Impala,[27] a massively parallel processing (MPP) SQL engine that can access HDFS and HBase data natively. This allows both large-scale batch processing using Hive and interactive queries using Impala to be performed on a common data–metadata storage architecture. Like Presto, Spark and Impala replace the map–reduce algorithm with their own optimized task schedulers.

[25]http://techblog.netflix.com/2014/10/using-presto-in-our-big-data-platform.html
[26]http://spark.apache.org
[27]http://www.cloudera.com/products/apache-hadoop/impala.html

CHAPTER 10

NoSQL Storage

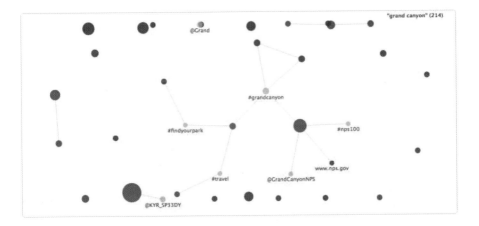

FIGURE 10.1 An affinity graph showing relationships between frequent tweets (green nodes), hashtags (yellow nodes), @-replies (orange nodes), and URLs (red notes) from Twitter discussions of the Grand Canyon

R ELATIONAL DATABASES, built on the relational model of Edgar Codd,[1] were proposed at IBM in the early 1970s. SQL (Structured Query Language, and initially called SEQUEL, for Structured English Query Language) was initially developed at IBM and first released commercially by Oracle (then Relational Software, Inc.) in 1979 as a special-purpose programming language to manage data stored in a relational database.

Various definitions of a relational database management system (RDBMS) exist, but they normally satisfy at least the following requirements.

- Data is structured as *relations*, tables consisting of rows (records) and columns (record attributes).

---

[1] A relational model of data for large shared data banks. Codd. *Communications of the ACM 13*, 6, 377–387, 1970.

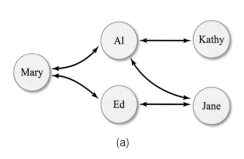

	(a)		Node	Friends

Node	Friends
Mary	{ Al, Ed }
Al	{ Jane, Kathy, Mary }
Ed	{ Jane, Kathy, Mary }
Kathy	{ Al }
Jane	{ Al, Ed }

(a)　　　　　　　　　　　　　(b)

FIGURE 10.2 Friend relationships: (a) friend relationship graph; (b) friend relationship RDBMS table

- Relational operators are provided to manipulate the tabular data.

The first commercially available RDBMS was Oracle, released in 1979. IBM followed with SQL/DS in 1981, and DB2 in 1983. Numerous RDBMSs are currently available, including Oracle Database and MySQL from Oracle, DB2 and Informix from IBM, SQL Server from Microsoft, and Sybase from SAP.

Relational databases are a powerful tool for storing structured numeric and string data. However, they are not always well-suited for other types of information. Recently, new database models have been proposed to store text documents, images, and node–link graphs. Our focus here is on graph databases and document databases. We will discuss how data is structured on external storage to support efficient retrieval in each domain.

10.1　GRAPH DATABASES

Various *graph databases* have been specifically designed to store and retrieve a graph's nodes and links. Examples include DEX, which was released in 2008, Hyper-GraphDB, which was presented at the International Workshop on Graph Databases in 2010, and Neo4j, which was released in 2010.

Graph databases claim a number of advantages over relational databases; in particular, they may be faster for associative datasets that represent discrete, independent entities and relationships between those entities. These correspond to a set of items and a set of relationships between the items, or in graph terms, a set of nodes and a set of links between pairs of nodes. Native graph databases implement this concept with *index-free adjacency*, where each node has explicit references to its adjacent nodes, and does not need to use an index to find them.

10.1.1　Neo4j

Although there are many ways to structure graph relationships in an RDBMS, one obvious approach is a table with a unique node ID as its key, and a list of nodes it links to as its value (Figure 10.2b).

property store

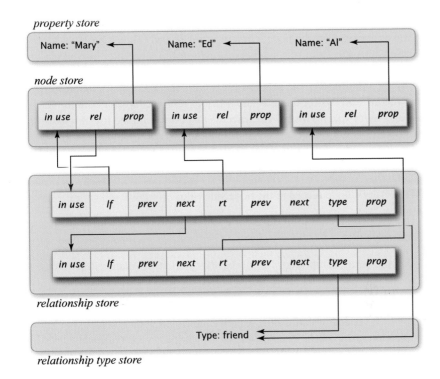

FIGURE 10.3 An overview of the property, node, relationship, and relationship type store files used to manage Mary's friend relationships

Consider finding friends of Mary. This is easy, since a direct lookup in the relational table returns the list { Al, Ed }. What if we wanted to find the friends of the friends of Mary? This is more expensive, since we need to perform a JOIN on Mary's friends to find *their* friends, producing the result { Mary, Jane }. In a more complicated graph friend-of-friend paths could be much longer. The farther we try to traverse from our starting node Mary, the more recursive JOIN operations we need to apply, and the more expensive the operation becomes. This is sometimes referred to by graph database researchers as a JOIN bomb. It highlights the point that path traversal in a graph stored in an RDBMS can be expensive.

A graph database replaces a relationship index table with node records that explicitly reference their relationships and local neighbor nodes. We describe the approach used by Neo4j,[2] although other graph databases use similar strategies. Neo4j stores data for different parts of the graph—nodes, relationships, properties, and so on—in different *store files* to improve graph traversal performance.

The node store file maintains node records: fixed-length records containing an

---

[2]*Graph Databases.* Robinson, Webber, and Eifrem, O'Reilly Media, Inc., Sebastopol, CA, 2013.

in-use flag, the ID (stored as an index) of the node's first relationship record, and the ID (stored as an index) of the node's first property record (Figure 10.3).

Relationships are stored in a separate relationship store file, also made up of fixed-length records containing an in-use flag, IDs of the start and end nodes connected by the relationship, the relationship's type, the ID of the relationship's first property record, and IDs for the next and previous relationships for both the start and end nodes (Figure 10.3). This provides access to a doubly linked list of relationships for both nodes, which can be traversed to search for target relationships and associated participant nodes. For example, suppose we wanted to find the friends of Mary's friends.

1. Find Mary's node record in the node store file.

2. Use the node record to retrieve Mary's first relationship record in the relationship store file.

3. Use the relationship record to access Mary's list of relationships. Walk forwards and backwards along this list, looking for "friend" relationships (Figure 10.3).

4. For each "friend" relationship, identify the friend node connected to Mary by this relationship.

5. For each friend node, apply similar processing to find *their* friend nodes. Combine and return these nodes as the friend-of-friend results.

Notice that, unlike in the RDBMS example, searching for longer friend-of-friend paths does not cause us to iterate recursively over the entire set of relationships. Index-free adjacency allows us to directly identify Mary's friends, then look only at those friends' relationships to determine the friend-of-friend results.

10.1.2 Caching

Neo4j uses two separate caches to improve IO performance: a lower-level *filesystem cache* for efficient block writes, and a higher-level *object cache* for efficient random-access reads.

The file system cache is made up of multiple *file buffer* caches, one per store file. Because the store files use fixed-length records, Neo4j can logically divide a store file into a sequence of fixed-sized windows, each of which holds multiple records. Every window records a hit count h, the number of requests for data in the window. For a set of n record requests, if the hit ratio h_i/n for uncached window i exceeds the miss ratio $1 - h_j/n$ for a cached window j, window j is evicted from its file buffer cache and replaced with window i.

The object cache stores nodes, relationships, and properties to support efficient in-memory graph traversal. Node objects contain both their relationships and their properties. Relationship objects contain only their properties. Node objects group their relationships by relationship type, allowing fast lookup based on this characteristic (e.g., searching for "friend" relationships). Cached objects are populated in a lazy fashion. For example, a node's relationships aren't loaded until a program attempts to access the relationships. A similar strategy is used for node and relationship properties. Objects are maintained in the cache using a least recently used (LRU) policy.

10.1.3 Query Languages

Since graph databases are not built on relational algebra, they cannot be queried with standard SQL. Many graph database query languages have been proposed, both in the research community and as an implementation in graph database systems.[3] For example, Neo4j supports Gremlin, a graph traversal language, and Cypher, a declarative query and manipulation language.

Gremlin communicates using the Blueprints interface, which Neo4j supports. Gremlin uses the concept of traversals to move from a starting location to (possibly multiple) ending locations in a graph. For example, a graph G containing information about Greek gods could be queried to find Heracles'[4] grandfather as follows.

```
G.V( 'name', 'hercules' ).out( 'father' ).out( 'father' ).name
==> cronus
G.V( 'name', 'hercules' ).out( 'father' ).out( 'father' ).roman_name
==> saturn
```

This query (1) finds the node in G with a name property of hercules; (2) traverses the outgoing edge with a property of father to find Heracles's father; (3) traverses the outgoing edge with a property of father to find Heracles's grandfather; and (4) returns the name property of the grandfather node.

Queries that evaluate along multiple branches are also possible, for example, to determine the lineage of Heracles' mother and father, or to determine the enemies that Heracles defeated.

```
G.V( 'name', 'hercules' ).out( 'mother', 'father' ).type
==> god
==> human
G.V( 'name', 'hercules' ).out( 'battled' ).map
==> { name=nemean, type=monster }
==> { name=hydra, type=monster }
==> { name=cerberus, type=monster }
```

Cypher is being developed as part of the Neo4j system, with a specific focus on readability and accessibility. Because of its declarative nature, Cypher allows you to define *what* to retrieve from a graph, but does not require you to describe *how* to

---

[3] A comparison of current graph database models. Angles. *Proceedings of the 28th International Conference on Data Engineering Workshops (ICDEW 2012)*, Arlington, VA, pp. 171–177, 2012.
[4] Known in Roman mythology as Hercules.

retrieve it. This has advantages and disadvantages. It often makes programming in Cypher simpler and easier to understand. However, it restricts the ability to optimize the order of traversals in a query. For complex queries, this can sometimes lead to significantly slower execution.

Similar to SQL, Cypher queries are built up as a set of chained clauses using operations like MATCH, WHERE, and RETURN. Cypher provides an online tutorial environment with a simple database containing information from the Matrix movies.[5]

- Movie nodes have title and year properties,

- Actor nodes have a name property, and

- ACTS_IN relationships connect an actor to a movie, and have a role property.

Basic queries use the MATCH operator. For example, the following queries return nodes for the movie "The Matrix" and the actor "Carrie-Anne Moss."

```
MATCH (m:Movie {title:"The Matrix"}) RETURN m;
(0:Movie {title:"The Matrix", year:"1999-03-31"})

MATCH (a:Actor {name:"Carrie-Anne Moss"}) RETURN a;
(5:Actor {name:"Carrie-Anne Moss"})
```

Relationships in the graph are specified with a -[type]-> or <-[type]- notation, depending on the relationship's direction. For example, this query returns all the actors and their roles in the movie "The Matrix."

```
MATCH (Movie {title:"The Matrix"})<-[r:ACTS_IN]-(a) RETURN r.role, a;
Neo (3:Actor {name:"Keanu Reeves"})
Morpheus (4:Actor {name:"Laurence Fishburn"})
Trinity (5:Actor {name:"Carrie-Anne Moss"})
```

The same approach could be used to find all the movies actor Keanu Reeves plays in, but with the direction of the ACT_IN relationship reversed.

```
MATCH (Actor {name:"Keanu Reeves"})-[ACTS_IN]->(m) RETURN m;
(0:Movie {title:"The Matrix", year:"1999-03-31"})
(1:Movie {title:"The Matrix Reloaded", year:"2003-05-07"})
(2:Movie {title:"The Matrix Revolutions", year:"2003-10-27"})
```

10.2 DOCUMENT DATABASES

Document databases are designed to store unstructured data. The term "unstructured data" is a bit of a misnomer. It describes data like text, audio, image, video, and so on. Clearly, these types of data have structure, but it's not the type of structure that's always easy to integrate into relational tables in an RDBMS.

As mentioned above, one common type of unstructured data is text: email, social media posts, documents, web pages, and so on. A document database makes it easy

---

[5] www.neo4j.org/learn/cypher

to mix this data together in a single table, or *collection*, without the need to formally define the table's structure. The document database sees each document as a "blob" of bytes, usually with associated properties that allow us to identify and retrieve individual documents.

Document databases are still being explored.[6] Various advantages are often cited, in particular, the ability for document databases to scale while maintaining good performance beyond what a typical RDBMS can support, and the ability to store different types of data without the need for a database schema.

Despite what some people may claim, it's not clear that document databases are a replacement for RDBMSs, or even a good alternative in many cases.[7] In spite of this, document databases like MongoDB, CouchDB, Cassandra, and Project Voldemort are becoming popular, so it's an interesting exercise to study *how* they store their data.

10.2.1 SQL Versus NoSQL

Understanding some of the major differences between SQL and NoSQL databases will help you to determine when to use one versus the other.[8] As mentioned above, proponents of NoSQL approaches highlight a number of advantages. First, NoSQL databases are *schemaless*. This means two documents can have different fields, or common fields that store different types of data.

```
var artists = [
  { first: "David", last: "Bowie", age: "sixty-six" },
  { first: "John", middle: "Winston", last: "Lennon", age: 30 }
];
```

Second, NoSQL databases like Mongo have built-in support for replication and sharding, allowing them to scale more easily than traditional SQL databases.

SQL approaches have their own advantages. First, they are built on a long history of research and technology, meaning they are purpose-built and proven in high capacity, production environments. SQL databases support fault tolerance and provide built-in journaling and transaction management to guarantee database consistency.

Second, because they are relational, SQL databases support JOINs. This allows columns from two or more tables to be combined based on a common field.

We could use an inner JOIN command in SQL to combine the order and customer tables into a new relational result that replaces CustID with each customer's name.

```
SELECT Orders.OrderID, Customers.Name, Orders.Date
FROM Orders
INNER JOIN Customers
ON Orders.CustID = Customers.CustID
```

Document databases like MongoDB do not provide a JOIN operation, since col-

---

[6] http://markedaspertinent.wordpress.com/2009/08/04/nosql-if-only-it-was-that-easy/

[7] http://www.sarahmei.com/blog/2013/11/11/why-you-should-never-use-mongodb/

[8] http://www.codemag.com/Article/1309051

TABLE 10.1 Three relational tables containing various information about customer orders

OrderID	CustID	Date
10308	2	2013-11-09
10309	4	2013-11-10
10310	1	2013-11-12
10310	2	2013-11-14

CustID	Name	Country
1	Steven Spielberg	USA
2	Rob Ford	Canada
3	Haruki Murakami	Japan
4	Sachin Tendulkar	India

OrderID	Name	Date
10308	Rob Ford	2013-11-09
10309	Sachin Tendulkar	2013-11-10
10310	Steven Spielberg	2013-11-12
10310	Rob Ford	2013-11-14

lections without a fixed schema won't necessarily have a common column to join over.

10.2.2 MongoDB

Our storage discussion focuses on MongoDB's approach to managing data.[9] A Mongo database is made up of one or more *collections*, each of which contains one or more *documents*. These are analogous to tables and records in an RDBMS.

On disk, a Mongo database is made up of a database namespace file db.ns, and two or more data storage files db.0, db.1,..., db.k. A *namespace* refers to a specific collection in a database, for example, db.artists. Each data storage file doubles in size, with db.0 occupying 64MB, db.1 occupying 128MB, and db.k occupying 64^{k+1}MB, up to a maximum size of 2GB. To improve performance, whenever the first write happens in db.k, Mongo preallocates db.k+1. db.ns and all data storage files are mapped into virtual memory, allowing Mongo to follow pointer references to access data.[10]

db.ns contains a large hash table with entries for each namespace—each collection in the database—and for any index files created over the database. A collection's hash entry includes the collection's name, statistics on its data, the offsets of its first and last data extents, and a set of availability lists (Figure 10.4). When a document is deleted, the resulting hole is placed on one of the availability lists. Mongo maintains

---

[9]http://www.mongodb.com/presentations/mongodbs-storage-engine-bit-bit

[10]Mapping database files into virtual memory is why Mongo recommends using a 64-bit server, which provides $\approx$127TB of virtual memory, versus $\approx$2.5GB on a 32-bit server.

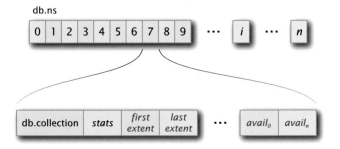

FIGURE 10.4 Mongo's namespace file, a hash table whose entries describe collections and indices; entries in red are disk offsets into a db.k file or an availability list file

multiple lists, each containing blocks of different fixed power-of-two sizes from 32 bytes to 8MB. To allocate a new document, Mongo finds the smallest available block larger than the new document, and assigns space such that both the document and any remaining hole(s) match the availability list block sizes.

Each collection is made up of a doubly linked list of extents, contiguous regions in a file that are owned by namespace db.k. db.ns identifies the location of the first and last extents in db.k. Each extent contains a header and a doubly linked list of documents, which hold the raw data stored in the database. An extent's header includes its location on disk, references to the next and previous extents, references to the first and last documents in the extent, and the size of the extent (Figure 10.5). The extent's data region holds the linked list of documents.

You can think of extents as the "backbone" of the database, allowing Mongo to store documents within each extent. A document includes a header with a size, a reference to its parent extent, references to the next and previous documents in the document list, and a data region with the document's raw data. Documents are stored in BSON (binary JSON) format. This allows them to contain internal structure that's accessible through Mongo's query APIs. For example, artist documents from our previous example could be searched by last name.

```
db.artists.find( { last: "Bowie" } );
```

Since each document is variable length, it's important to understand how Mongo allocates space within an extent. Initially, each new document in a collection is assigned a block from an availability list using a "best fit" strategy. If a document is later updated to be larger, it may need to be moved to a bigger block. When this happens, a *padding factor* p is incremented, and all new block requests are defined to be the size of a document times the padding factor. The intuition here is that, since documents are being updated to be larger, we should pad their blocks to try to allow the update to happen in place. If an update occurs that doesn't require a document to be moved, the padding factor is decremented. p is local to a collection, and its range is always $1.0 \le p \le 2.0$.

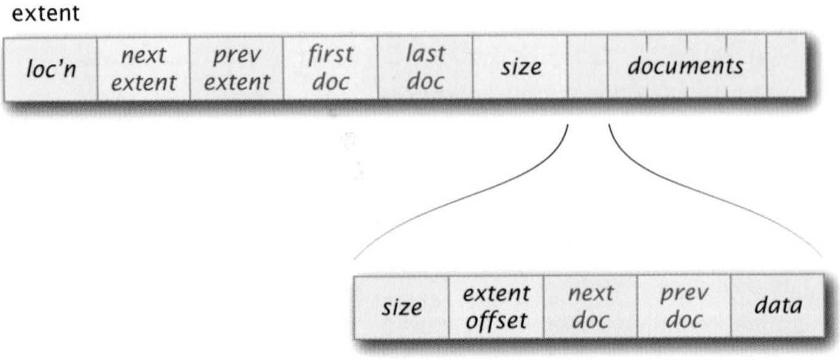

FIGURE 10.5 Mongo's extent structure, with header information and a collection of document records in the data region; each document record has its own header plus the raw data that makes up the document; entries in red are disk offsets and entries in green are memory offsets within the extent

Given how Mongo allocates blocks and moves documents, it's possible for significant external fragmentation to occur within a database. To address this, Mongo provides a command to compact a collection and reset its availability lists. Mongo databases are normally replicated as a primary and one or more secondary mirrors. The secondary collections are compacted one by one, a compacted secondary is promoted to replace the primary, then the (old) primary is compacted. This allows compacting without taking the database offline.

10.2.3 Indexing

Like an RDBMS, Mongo constructs index files to support efficient searching. Mongo uses a B-tree with 8KB buckets to maintain its indices. Mongo assumes the majority of the operations will be insertion, however, so it performs a 90–10 split when a bucket overflows, keeping 90% of the data in the original bucket and moving 10% to the new bucket. This ensures that most buckets are mostly full, providing better space utilization.

Because Mongo doesn't use a schema like a traditional RDBMS, it can't use a fixed-length size for key values in the B-tree node. This can impact the performance of searching the key list within a node. To address this, each node's data is structured into fixed-length and variable-length regions (Figure 10.6).

- a fixed-length reference to the parent node,

- a fixed-length reference to the rightmost child subtree beneath this node,

- an array of fixed-length *key nodes*, each of which holds a reference to a left child subtree, a reference to the document attached to this key, and a reference to a key value stored in the node's *key object* list, and

- a variable-length key object list containing the actual values for each key stored in the node.

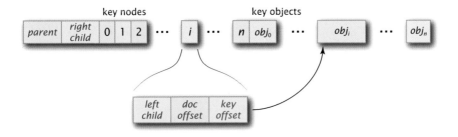

FIGURE 10.6 Mongo's B-tree node: (1) a fixed-length section holding parent reference, rightmost subtree reference, and key nodes with references to each key's left subtree, its document, and its value; and (2) a variable-length key object section holding actual key values; entries in red are disk offsets and entries in green are memory offsets within the node

The key node array is sorted by key value. Since it's made up of fixed-length records, it can be searched fairly rapidly. The actual key values are stored in unsorted order in the key object list, a block of memory in the B-tree node divided into variable-length blocks to hold each key's value.

10.2.4 Query Languages

Most NoSQL databases include their own command shells, APIs, or programming languages to provide access to a database's structure and contents. For example, MongoDB includes the mongo JavaScript shell for interactive manipulation of its databases.[11] This is useful, because it lets you combine JavaScript with Mongo operations to build scripts to manipulate a database. Mongo models these interactions around create, read, update, and delete (CRUD) operations.[12] For example, the following JavaScript creates a movie collection in the current database, then stores three "documents" within the collection.

```
> doc = [
    { "name": "The Matrix", "year": "1993-02-31" },
    { "name": "The Matrix Reloaded", "year": "2003-05-07" },
    { "name": "The Matrix Revolutions", "year": "2003-10-27" }
  ];
> for( i = 0; i < doc.length; i++ ) {
    db.movie.save( doc[ i ] );
  }
```

---

[11]http://try.mongodb.org/
[12]http://docs.mongodb.org/manual/crud/

Because the shell accepts JavaScript, documents can be specified as JSON objects. Recall that these are then stored as BSON in the collection's storage files. Additional commands allow for querying, updating, deleting, and so on.

```
> db.movie.find( { "name": "The Matrix" } )
{ "name": "The Matrix", "year": "1993-02-31" }

> db.movie.update( { "name": "The Matrix" }, { "$set": { "year":
"1993-03-21" }} );
> db.movie.find( { "name": "The Matrix" } )
{ "name": "The Matrix", "year": "1993-03-31" }

> db.movie.remove( { "name": "The Matrix Revolutions" } )
> db.movie.find()
{ "name": "The Matrix", "year": "1993-03-31" } }
{ "name": "The Matrix Reloaded", "year": "2003-05-07" } }
```

Mongo also provides APIs to numerous external languages, including C and C++, Java, PHP, Python, and Ruby.[13]

---

[13]http://api.mongodb.org

Order Notation

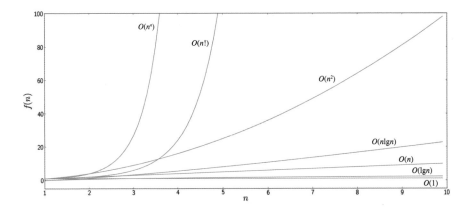

FIGURE A.1 The asymptotic growth for different order notation function classes

ORDER NOTATION is a mathematical method for bounding the performance of an algorithm as its size grows without bound. It allows us to define and compare an algorithm's performance in a way that is free from uncontrolled influences like machine load, implementation efficiency, and so on.

We provide a brief review of order notation, to ensure a common understanding of its terminology. This is not meant to be a comprehensive discussion of algorithm analysis. In fact, although we use order notation throughout this course, we are often "flexible," for example, by only measuring certain parts of an algorithm (e.g., seek times) when we know these operations account for the vast majority of the algorithm's execution time.

A.1 Θ-NOTATION

Θ-notation provides asymptotic upper and lower bounds on the efficiency of an algorithm for a particular input size n.

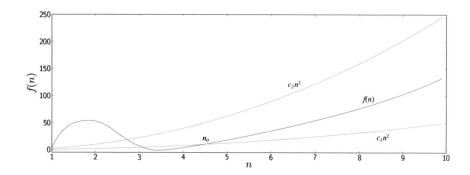

FIGURE A.2 A function $f(n)$ bounded above and below by n^2, showing $f(n) = \Theta(n^2)$

$$\Theta(g(n)) = \{\, f(n) : \exists\, c_1, c_2, n_0 > 0 \mid$$
$$0 \le c_1\, g(n) \le f(n) \le c_2\, g(n) \ \forall\, n \ge n_0 \,\} \tag{A.1}$$

Saying $f(n) = \Theta(g(n))$ means $f(n) \in \Theta(g(n))$. For example, suppose $f(n) = \frac{1}{2}n^2 - 3n$. We claim $f(n) = \Theta(n^2)$. To prove this, we must find $c_1, c_2, n_0 > 0$ satisfying the inequality

$$c_1\, n^2 \le \frac{1}{2}n^2 - 3n \le c_2\, n^2 \tag{A.2}$$

Dividing through by n^2 yields

$$c_1 \le \frac{1}{2} - \frac{3}{n} \le c_2 \tag{A.3}$$

If $c_2 = \frac{1}{2}$, Eq. A.3 is true $\forall\, n \ge 1$. Similarly, if $c_1 = \frac{1}{14}$, Eq. A.3 is true $\forall\, n \ge 7$ (for $n = 7$, $\frac{1}{2} - \frac{3}{7} = \frac{1}{14}$). $\therefore$ Eq. A.3 is true for $c_1 = \frac{1}{14}, c_2 = \frac{1}{2}, n_0 = 7$.

Note that there are many possible c_1, c_2, n_0 that we could have chosen. The key is that there must exist at least *one* set of c_1, c_2, n_0 values that satisfy our constraints.

A.2 O-NOTATION

O-notation provides an asymptotic upper bound on the efficiency of an algorithm for a particular input size n.

$$O(g(n)) = \{\, f(n) : \exists\, c, n_0 > 0 \mid$$
$$0 \le f(n) \le c\, g(n) \ \forall\, n \ge n_0 \,\} \tag{A.4}$$

Note that if an algorithm is $f(n) = O(n)$, then by definition it is also $O(n^2)$,

$O(n^3)$, and so on. When we say $f(n) = O(g(n))$, we normally try to choose $g(n)$ to define a *tight* upper bound. Also, $f(n) = \Theta(g(n)) \implies f(n) = O(g(n))$, that is, $\Theta(g(n)) \subset O(g(n))$.

If $f(n) = O(g(n))$ for worst-case input, then $f(n) = O(g(n))$ for any input. This is not true for Θ-notation. If $f(n) = \Theta(g(n))$ for worst-case input, it does not imply that $f(n) = \Theta(g(n))$ for all input. Certain inputs may provide performance better than $g(n)$.

A.3 Ω-NOTATION

Ω-notation provides an asymptotic lower bound on the efficiency of an algorithm for a particular input size n.

$$\Omega(g(n)) = \{\, f(n) : \exists\, c, n_0 > 0 \mid$$
$$0 \le c\, g(n) \le f(n) \;\forall\, n \ge n_0 \,\} \tag{A.5}$$

As with O-notation, $f(n) = \Theta(g(n)) \implies f(n) = \Omega(g(n))$, that is, $\Theta(g(n)) \subset \Omega(g(n))$. In fact, $f(n) = \Theta(g(n)) \iff f(n) = O(g(n))$ and $f(n) = \Omega(g(n))$.

A.4 INSERTION SORT

As a practical example, consider measuring the performance of insertion sort, where performance is represented by the number of statements executed within the sort. An insertion sort of an array $A[1 \ldots n]$ of size n pushes each element $A[2]$ through $A[n]$ to the left, into its sorted position in the front of the array.

```
insertion_sort(A, n)
Input: A[ ], array of integers to sort; n, size of A

for j = 2 to n                              n
do
    key = A[ j ]                            n − 1
    i = j - 1                               n − 1
    while i ≥ 1 and A[ i ] > key            ∑_{j=1}^{n-1} t_j
    do
        A[ i + 1 ] = A[ i ]                 ∑_{j=1}^{n-1}(t_j − 1)
        i--                                 ∑_{j=1}^{n-1}(t_j − 1)
    end
    A[ i + 1 ] = key                        n − 1
end
```

The numbers to the right of each statement represent a count of the number of times the statement is executed. These are based either on n, the size of the array, or t_j, the number of executions on the j-th pass through the inner while loop. We can simply sum up the execution counts to get an overall cost $T(n)$ for an insertion sort on an array of size n.

$$T(n) = n + (n-1) + (n-1) + \sum_{j=1}^{n-1} t_j + \sum_{j=1}^{n-1} (t_j - 1) + \sum_{j=1}^{n-1} (t_j - 1) + (n-1) \quad \text{(A.6)}$$

Suppose we encountered the best-case input, an already sorted array. In this case $t_j = 1 \ \forall \ j = 2, \ldots, n$, since we never enter the body of the while loop (because $A[i] \leq A[i+1] \ \forall \ i = 1, \ldots, n$). The cost in this case is

$$T(n) = n + (n-1) + (n-1) + \sum_{j=1}^{n-1} 1 + \sum_{j=1}^{n-1} 0 + \sum_{j=1}^{n-1} 0 + (n-1)$$

$$= 4n - 3 + \sum_{i=1}^{n-1} 1 \quad \text{(A.7)}$$

$$= 4n - 3 + (n-1)$$

$$= 5n - 4$$

This shows best-case performance of $\Theta(n)$.

In the worst case, A is reverse sorted, and the while loop must be executed j times for the j-th pass through the for loop. This means $t_j = j$. To compute total cost, we need to solve for $\sum_{j=1}^{n-1} j$ and $\sum_{j=1}^{n-1} (j-1)$. Recall

$$\sum_{j=1}^{n-1} j = \frac{n(n-1)}{2} \quad \text{(A.8)}$$

$$\sum_{j=1}^{n-1} (j-1) = \sum_{j=1}^{n-1} j - \sum_{j=1}^{n-1} 1$$

$$= \frac{n(n-1)}{2} - (n-1) \quad \text{(A.9)}$$

$$= \frac{n(n-3)}{2} + 1$$

Given these sums, the total cost $T(n)$ can be calculated.

$$T(n) = n + (n-1) + (n-1) + \frac{n(n-1)}{2} + \frac{n(n-3)}{2} + 1 + \frac{n(n-3)}{2} + 1 + (n-1)$$

$$= \frac{3n^2}{2} + \frac{n}{2} - 1 \quad \text{(A.10)}$$

The n^2 term in $T(n)$ dominates all other terms as n grows large, so $T(n) = \Theta(n^2)$ for worst-case input. Given this, the worst case $T(n) = O(n^2)$ as well, so we often say that insertion sort runs in $O(n^2)$ in the worst case.

In the average case, we could argue that each element in A would need to move about half the distance from its starting position to the front of A. This means $t_j = \frac{j}{2}$.

Although we don't provide the details here, calculating $T(n)$ shows that, on average, it is also $\Theta(n^2)$.

In spite of insertion sort's $\Theta(n^2)$ average case performance, it is one of the fastest sorts, in absolute terms, for small arrays. For example, insertion sort is often used in Quicksort to sort a partition once its length falls below some predefined threshold, because this is faster than recursively finishing the Quicksort.

A.5 SHELL SORT

As an example of a more sophisticated analysis of time complexity, consider Shell sort, proposed by Donald L. Shell in 1956. Similar to insertion sort, it takes an element $A[m]$ and compares it to previous elements to push it into place. Rather than comparing with the neighbor $A[m-1]$ immediately to the left, however, we use larger increments of size $h > 1$ to generate sorted subgroups. We then slowly decrease the increment size until $h = 1$, at which point the array will be sorted. Because of this, Shell sort is sometimes referred to as diminishing insertion sort.

As an example, suppose we have a 16-element array $A[0 \ldots 15]$ and we divide it into eight groups with elements separated by an $h = 8$ increment.

$$(A_0, A_8), \ (A_1, A_9), \ \ldots \ (A_7, A_{15}) \tag{A.11}$$

We sort each group independently using insertion sort. After we're done, we know that $A_0 \le A_8$, $A_1 \le A_9$, $\ldots A_7 \le A_{15}$. Now, we reduce the increment to $h = 4$, producing groups of size four.

$$(A_0, A_4, A_8, A_{12}), \ \ldots \ (A_3, A_7, A_{11}, A_{15}) \tag{A.12}$$

We again apply insertion sort to sort each of these subgroups. Notice something important, however. Because of the previous step with $h = 8$, each subgroup is already partially sorted. For example, in the first subgroup we know that $A_0 \le A_8$ and $A_4 \le A_{12}$. This means that when we push A_8 into place, it may swap with A_4, but it will never swap with A_0 because it's already "in place" relative to A_0.

This shows that the previous sorting step reduces the number of comparisons needed to complete the current sorting step. We continue to sort subgroups and reduce the increment down to $h = 1$. At this point, the array is already almost sorted due to the $h = 2$ sort step (that is, we know $A_0 \le A_2$, $A_1 \le A_3$, $\ldots A_{13} \le A_{15}$). This means the final insertion sort should run much faster than an insertion sort on the array in its original configuration.

The question for Shell sort is: Do the multiple insertion sorts with increments $h > 1$ cost more than the savings we achieve when we perform the final insertion sort with $h = 1$? It turns out we save more on the final insertion sort than it costs to configure the array for that sort, so Shell sort is, on average, faster than insertion sort.

Consider the simplest generalization. We will perform two steps, one with $h = 2$ followed by another with $h = 1$. After the $h = 2$ sort, we have a 2-sorted array. Let's analyze the performance of the final step, placing the 2-sorted array into order.

How many permutations p are there for n values such that $A_i \leq A_{i+2} \; \forall \, 0 \leq i \leq n - 3$, that is, n values that are 2-sorted? p is n choose $^n/_2$, or more formally:

$$p = \binom{n}{\lfloor \frac{n}{2} \rfloor} \tag{A.13}$$

Each permutation is equally likely. For any permutation, how many swaps are required to go from 2-sorted to "in order"? Let's define A_n to be the total number of swaps needed to put all possible 2-sorted lists with n values in order. We can hand-compute this value for small n.

n	2-Sorted Permutations	Swaps
$n = 1$	$\{1\}$	$A_1 = 0$
$n = 2$	$\{12, 21\}$	$A_2 = 0 + 1 = 1$
$n = 3$	$\{123, 132, 213\}$	$A_3 = 0 + 1 + 1 = 2$
$n = 4$	$\{1324, 1234, 1243, 2134, 2143, 3142\}$	$A_4 = 1 + 0 + 1 + 1 + 2 + 3 = 8$

It turns out that the total number of swaps for a given n is $A_n = \lfloor {}^n/_2 \rfloor \, 2^{n-2}$. Therefore, the average number of swaps for any given permutation is

$$A_n = \frac{\lfloor \frac{n}{2} \rfloor 2^{n-2}}{p} = \frac{\lfloor \frac{n}{2} \rfloor 2^{n-2}}{\binom{n}{\lfloor \frac{n}{2} \rfloor}} \approx 0.15 n^{3/2} \tag{A.14}$$

This means the final insertion sort on a 2-sorted array runs in $O(n^{3/2})$ time, which is better than $O(n^2)$. This is the savings we achieve if we first perform an insertion sort with $h = 2$ to produce the 2-sorted array.

How much does it cost to do the $h = 2$ sort step? We will not provide the details here, but it can be proven that 2-sorting an array of size n requires $O(n^{5/3})$ time. This means the overall cost of Shell sort with two steps is $O(n^{5/3})$, which is still better than insertion sort's $O(n^2)$ average case performance.

More steps that use larger increments add some additional improvements. The results will always fall somewhere between $O(n^{5/3})$ and $O(n^2)$, that is, Shell sort is never worse than insertion sort, but a careful choice of h must be made at each step to see improved performance.

What is a "good" h? No theoretical answer to this question is known, but empirical evidence suggests

$$h_1 = 1$$
$$h_2 = 3h_1 + 1 = 4$$
$$h_3 = 3h_2 + 1 = 13 \tag{A.15}$$
$$\cdots$$
$$h_{s+1} = 3h_s + 1$$

We stop when $h_{s+2} \geq n$. A little math shows this is equivalent to $h_s = \lceil \frac{n-4}{9} \rceil$. This is the increment we use for the first insertion sort step.

Assignment 1 : Search

FIGURE B.1 In-memory versus disk-based searching

THE GOALS for this assignment are two-fold:

1. To introduce you to random-access file I/O in UNIX using C.

2. To investigate time efficiency issues associated with in-memory versus disk-based searching strategies.

This assignment uses two "lists" of integer values: a key list and a seek list. The key list is a collection of integers $K = (k_0, \ldots, k_{n-1})$ representing n keys for a hypothetical database. The seek list is a separate collection of integers $S = (s_0, \ldots, s_{m-1})$ representing a sequence of m requests for keys to be retrieved from the database.

You will implement two different search strategies to try to locate each s_i from the seek list:

1. **Linear search.** A sequential search of K for a key that matches the current seek value s_i.

2. **Binary search.** A binary search through a sorted list of keys K for a key that matches the current seek value s_i. The fact that the keys are sorted allows approximately half the remaining keys to be ignored from consideration during each step of the search.

Each of the two searches (linear and binary) will be performed in two different environments. In the first, the key list K will be held completely in memory. In the second, individual elements $k_i \in K$ will read from disk as they are needed.

B.1 KEY AND SEEK LISTS

The key and seek lists are provided to you as binary files. Each binary file contains a sequence of integer values stored one after another in order. You can download examples of both files from the Supplemental Material section of this appendix.

Be sure to capture these files as binary data. The example file sizes should be 20,000 bytes for the key file, and 40,000 bytes for the seek file. For simplicity, the remainder of the assignment refers only to key.db and seek.db.

Note. Apart from holding integers, you cannot make any assumptions about the size or the content of the key and seek files we will use to test your program.

B.2 PROGRAM EXECUTION

Your program will be named assn_1 and it will run from the command line. Three command line arguments will be specified: a search mode, the name of the key file, and the name of the seek file.

```
assn_1 search-mode keyfile-name seekfile-name
```

Your program must support four different search modes.

1. --mem-lin   Read the key file into memory. Perform a linear search for each seek element s_i in the seek file.

2. --mem-bin   Read the key file into memory. Perform a binary search for each seek element s_i in the seek file.

3. --disk-lin   Read each k_i from the key file as it is needed. Perform a linear search for each seek element s_i in the seek file.

4. --disk-bin   Read each k_i from the key file as it is needed. Perform a binary search for each seek element s_i in the seek file.

For example, executing your program as follows

```
assn_1 --disk-lin key.db seek.db
```

would search for each element in seek.db using an on-disk linear search within key.db.

B.3   IN-MEMORY SEQUENTIAL SEARCH

If your program sees the search mode --mem-lin, it will implement an in-memory sequential search of the key list stored in key.db. The program should perform the following steps.

1. Open and read seek.db into an appropriately sized integer array S.

2. Open and read key.db into an appropriately sized integer array K.

3. Create a third array of integers called hit of the same size as S. You will use this array to record whether each seek value S[i] exists in K or not.

4. For each S[i], search K sequentially from beginning to end for a matching key value. If S[i] is found in K, set hit[i]=1. If S[i] is not found in K, set hit[i]=0.

You must record how much time it takes to open and load key.db, and to then determine the presence or absence of each S[i]. This is the total cost of performing the necessary steps in an in-memory sequential search. Be sure to measure only the time needed for these two steps: loading key.db and searching K for each S[i]. Any other processing should not be included.

B.4   IN-MEMORY BINARY SEARCH

If your program sees the search mode --mem-bin, it will implement an in-memory binary search of the key list stored in key.db. The keys in key.db are stored in sorted order, so they can be read and searched directly. Your program should perform the following steps.

1. Open and read seek.db into an appropriately sized integer array S.

2. Open and read key.db into an appropriately sized integer array K.

3. Create a third array of integers called hit of the same size as S. You will use this array to record whether each seek value S[i] exists in K or not.

4. For each S[i], use a binary search on K to find a matching key value. If S[i] is found in K, set hit[i]=1. If S[i] is not found, set hit[i]=0.

You must record how much time it takes to open and load key.db, and to then determine the presence or absence of each S[i]. This is the total cost of performing the necessary steps in an in-memory binary search. Be sure to measure only the time needed for these two steps: loading key.db and searching K for each S[i]. Any other processing should not be included.

Recall. To perform a binary search for S[i] in an array K of size n, begin by comparing S[i] to K[$\frac{n}{2}$].

- If S[i] == K[$\frac{n}{2}$], the search succeeds.

- If $S[i] < K[\frac{n}{2}]$, recursively search the lower subarray $K[0] \dots K[\frac{n}{2} - 1]$ for $S[i]$.

- Otherwise, recursively search the upper subarray $K[\frac{n}{2}+1] \dots K[n-1]$ for $S[i]$.

Continue recursively searching for $S[i]$ and dividing the subarray until $S[i]$ is found, or until the size of the subarray to search is 0, indicating the search has failed.

B.5   ON-DISK SEQUENTIAL SEARCH

For on-disk search modes, you will not load key.db into an array in memory. Instead, you will search the file directly on disk.

If your program sees the search mode --disk-lin, it will implement an on-disk sequential search of the key list stored in key.db. The program should perform the following steps.

1. Open and read seek.db into an appropriately sized integer array S.

2. Open key.db for reading.

3. Create a second array of integers called hit of the same size as S. You will use this array to record whether each seek value $S[i]$ exists in key.db or not.

4. For each $S[i]$, search key.db sequentially from beginning to end for a matching key value by reading K_0 and comparing it to $S[i]$, reading K_1 and comparing it to $S[i]$, and so on. If $S[i]$ is found in key.db, set hit[i]=1. If $S[i]$ is not found in key.db, set hit[i]=0.

You must record how much time it takes to determine the presence or absence of each $S[i]$ in key.db. This is the total cost of performing the necessary steps in an on-disk sequential search. Be sure to measure only the time needed to search key.db for each $S[i]$. Any other processing should not be included.

Note. If you read past the end of a file in C, its EOF flag is set. Before you can perform any other operations on the file, you must reset the EOF flag. There are two ways to do this: (1) close and reopen the file; or (2) use the clearerr() function to clear the FILE stream's EOF and error bits.

B.6   ON-DISK BINARY SEARCH

If your program sees the search mode --disk-bin, it will implement an on-disk binary search of the key list stored in key.db. The keys in key.db are stored in sorted order, so they can be read and searched directly. The program should perform the following steps.

1. Open and read seek.db into an appropriately sized integer array S.

2. Open key.db for reading.

3. Create a second array of integers called hit of the same size as S. You will use this array to record whether each seek value S[i] exists in key.db or not.

4. For each S[i], use a binary search on key.db to find a matching key value. If S[i] is found in key.db, set hit[i]=1. If S[i] is not found in key.db, set hit[i]=0.

You must record how much time it takes to to determine the presence or absence of each S[i] in key.db. This is the total cost of performing the necessary steps in an on-disk binary search. Be sure to measure only the time needed to search key.db for each S[i]. Any other processing should not be included.

B.7   PROGRAMMING ENVIRONMENT

All programs must be written in C, and compiled to run on a Linux system.

B.7.1   Reading Binary Integers

C's built-in file operations allow you to easily read integer data stored in a binary file. For example, the following code snippet opens a binary integer file for input and reads three integers: the first integer in the file, the third integer from the start of the file, and the second integer from the end of the file.

```
#include <stdio.h>

FILE *inp;          /* Input file stream */
int   k1;           /* Keys to read */
int   k2;
int   k3;

inp = fopen( "\texttt{key.db}", "rb" );

fread( &k1, sizeof( int ), 1, inp )

fseek( inp, 2 * sizeof( int ), SEEK_SET );
fread( &k2, sizeof( int ), 1, inp )

fseek( inp, -2 * sizeof( int ), SEEK_END );
fread( &k3, sizeof( int ), 1, inp )
```

B.7.2 Measuring Time

The simplest way to measure execution time is to use gettimeofday() to query the current time at appropriate locations in your program.

```
#include <sys/time.h>

struct timeval tm;

gettimeofday( &tm, NULL );
printf( "Seconds: %d\n", tm.tv_sec );
printf( "Microseconds: %d\n", tm.tv_usec );
```

Comparing tv_sec and tv_usec for two timeval structs will allow you to measure the amount of time that's passed between two gettimeofday() calls.

B.7.3 Writing Results

Results for each key in S[i], and the total time needed to perform all searching, must be written to the console before your program terminates. The format of your output must conform to the following rules.

1. Print one line for each S[i] in the order it occurs in seek.db. The line must contain the value of S[i] padded to be twelve characters wide, a colon and a space, and the text Yes if hit[i]=1 (key found) or the text No if hit[i]=0 (key not found). The simplest way to do this is to use a printf statement, for example,

   ```
   printf( "%12d: Yes", \texttt{S[i]} );
   ```

2. Print the total execution time for your program's search operations as a single line with the label Time: followed by a space, and the execution time in seconds and microseconds. Assuming the execution time is stored in a timeval struct called exec_tm, you can use a printf statement to do this.

   ```
   printf( "Time: %ld.%06ld", exec_tm.tv_sec, exec_tm.tv_usec );
   ```

You can capture your program's results for further examination or validation by redirecting its output to an external file, for example, to a file called output.txt, as follows.

```
assn_1 --mem-lin key.db seek.db > output.txt
```

Your assignment will be run automatically, and the output it produces will be compared to known, correct output using diff. Because of this, **your output must conform to the above requirements exactly**. If it doesn't, diff will report your output as incorrect, and it will be marked accordingly.

B.8   SUPPLEMENTAL MATERIAL

In order to help you test your program, we provide an example key.db and seek.db, as well as the output that your program should generate when it processes these files.

- key.db, a binary integer key list file containing 5000 keys (http://go.ncsu.edu/big-data-assn1-key.db),

- seek.db, a binary integer seek list file containing 10000 keys (http://go.ncsu.edu/big-data-assn1-seek.db), and

- output.txt, the output your program should generate when it processes key.db and seek.db (http://go.ncsu.edu/big-data-assn1-output.txt).

You can use diff to compare output from your program to our output.txt file. Note that the final line containing the execution time most likely won't match, but if your program is running properly and your output is formatted correctly, all of the key searches should produce identical results.

Please remember, as emphasized previously, the files we're providing here are meant to serve as examples only. Apart from holding integers, **you cannot make any assumptions** about the size or the content of the key and seek files we will use to test your program.

B.9   HAND-IN REQUIREMENTS

Use the online assignment submission system to submit the following files:

- assn_1, a Linux executable of your finished assignment, and

- all associated source code files (these can be called anything you want).

There are four important requirements that your assignment must satisfy.

1. Your executable file must be named exactly as shown above. The program will be run and marked electronically using a script file, so using a different name means the executable will not be found, and subsequently will not be marked.

2. Your program must be compiled to run on a Linux system. If we cannot run your program, we will not be able to mark it, and we will be forced to assign you a grade of zero.

3. Your program must produce output that exactly matches the format described in the Writing Results section of this assignment. If it doesn't, it will not pass our automatic comparison to known, correct output.

4. You must submit your source code with your executable prior to the submission deadline. If you do not submit your source code, we cannot MOSS it to check for code similarity. Because of this, any assignment that does not include source code will be assigned a grade of zero.

Assignment 2: Indices

FIGURE C.1 In-memory indexing with availability lists

T HE GOALS for this assignment are three-fold:

1. To investigate the use of field delimiters and record sizes for field and record organization.

2. To build and maintain an in-memory primary key index to improve search efficiency.

3. To use an in-memory availability list to support the reallocation of space for records that are deleted.

C.1   STUDENT FILE

During this assignment you will build and maintain a simple file of student records. Each record will have four fields: SID (student ID), last (last name), first (first name), and `major` (program of study). Fields within a record will be variable length, and will be delimited by the | character. For example,

```
712412913|Ford|Rob|Phi
```

represents a student with an SID of 712412913, a `last` of Ford, a `first` of Rob, and a major of Phi (Rob Ford is minoring in Ethics).

SID is the primary key for a student record. This means every individual student record will have a unique SID.

Records will be variable length, and will be stored one after another in a binary data file. Each record will be preceded by an integer that defines the size of its corresponding record.

Note. Read the above description of the record size carefully! It is stored as binary data in a manner similar to how integer data were stored and retrieved in Assignment 1. It is not appropriate to store the size as an ASCII string. For example, if you wanted to read a record at file offset `off` in a file referenced through a `FILE` stream `fp`, it would be done as

```c
#include <stdio.h>

char *buf;      /* Buffer to hold student record */
FILE *fp;       /* Input/output stream */
long  rec_off;  /* Record offset */
int   rec_siz;  /* Record size, in characters */

/*  If student.db doesn't exist, create it, otherwise read
 *  its first record
 */

if ( ( fp = fopen( "student.db", "r+b" ) ) == NULL ) {
  fp = fopen( "student.db", "w+b" );

} else {
  rec_off = 0;
  fseek( fp, rec_off, SEEK_SET );
  fread( &rec_siz, sizeof( int ), 1, fp );

  buf = malloc( rec_siz );
  fread( buf, 1, rec_siz, fp );
}
```

Writing a new record uses a similar, reverse procedure. First, convert the record's body into a character buffer with fields delimited by |. Next, seek to the appropriate position in the student file and write an integer representing the buffer's size in bytes, in binary, using `fwrite()`. Finally, write the buffer to the file with `fwrite()`.

C.2   PROGRAM EXECUTION

Your program will be named `assn_2` and it will run from the command line. Two command line arguments will be specified: an availability list order, and the name of the student file.

```
assn_2 avail-list-order studentfile-name
```

Your program must support three different availability list ordering strategies.

1. `--first-fit`   Holes in the availability list will be saved in the order they are added to the list.

2. `--best-fit`   Holes in the availability list will be sorted in ascending order of hole size.

3. `--worst-fit`   Holes in the availability list will be sorted in descending order of hole size.

For example, executing your program as follows

```
assn_2 --best-fit student.db
```

would order holes in the availability list in ascending order of hole size, and would use student.db to store records.

Note. If you are asked to open an existing student file, you can assume the availability list order specified on the command line matches the order that was used when the student file was first created.

C.3   IN-MEMORY PRIMARY KEY INDEX

In order to improve search efficiency, a primary key index will be maintained in memory as a reference to each record stored in the file. For our records, SID will act as a primary key. This means each entry in your index will have a structure similar to

```
typedef struct {
  int key;    /* Record's key */
  long off;   /* Record's offset in file */
} index_S;
```

Index entries should be stored in a collection that supports direct access and dynamic expansion. One good choice is a dynamically expandable array. Index entries must be sorted in ascending key order, with the smallest key at the front of the index and the largest key at the end. This will allow you to use a binary search to find a target key in the index. The index will not be rebuilt every time the student file is reopened. Instead, it will be maintained in a persistent form on disk. As your program exits, you will write your index to disk, saving its contents in an index file. When you reopen the student file, you will load the corresponding index file, immediately reconstructing your primary key index. The index will have exactly the same state as before, and will be ready to use to access records in the student file.

You can use any format you want to store each key–offset pair in the index file. The simplest approach is to read and write the entire structure as binary data using fread() and fwrite(), for example,

```
#include <stdio.h>

typedef struct {
  int key;   /* Record's key */
  long off;  /* Record's offset in file */
} index_S;

FILE     *out;            /* Output file stream */
index_S  prim[ 50 ];  /* Primary key index */

out = fopen( "index.bin", "wb" );
fwrite( prim, sizeof( index_S ), 50, out );
fclose( out );
```

Note. To simplify your program, you can assume the primary key index will never need to store more than 10,000 key–offset pairs.

C.4   IN-MEMORY AVAILABILITY LIST

When records are deleted from the file, rather than closing the hole that forms (an expensive operation), we will simply record the size and the offset of the hole in an in-memory availability list. Each entry in your availability list will have a structure similar to

```
typedef struct {
  int siz;   /* Hole's size */
  long off;  /* Hole's offset in file */
} avail_S;
```

Note. To simplify your program, you can assume the availability list will never need to store more than 10,000 size–offset pairs.

The availability list will not be rebuilt every time the student file is reopened. Instead, similar to the primary index, it will be maintained in a persistent form on disk. As your program exits, you will write your availability list to disk, saving its contents in an availability list file. When you reopen the student file, you will load the corresponding availability list file, immediately reconstructing your availability list.

As noted above, you can assume a consistent availability list order for a given student file. In other words, if you are asked to open an existing student file, the availability list order specified on the command line will always match the order that was being used when the availability list was written to disk.

When new records are added, we will search the availability list for a hole that can hold the new record. If no hole exists, the record is appended to the end of the student file. The order of the entries in the availability list is defined by the availability order specified on the command line when your program is run.

C.4.1 First Fit

If your program sees the availability order `--first-fit`, it will order entries in the availability list in first in–first out order. New holes are appended to the end of the availability list. When a new record is added, a first-fit strategy is used to search from the front of the availability list until a hole is found that is large enough to hold the new record.

If the hole is larger than the new record, the left-over fragment is saved as a new hole at the end of the availability list.

C.4.2 Best Fit

If your program sees the availability order `--best-fit`, it will order entries in the availability list sorted in ascending order of hole size. New holes are inserted into the availability list in the proper sorted position. If multiple holes have the same size, the entries for these holes should be sorted in ascending order of hole offset.

When a new record is added, a best-fit strategy is used to search from the front of the availability list until a hole is found that is large enough to hold the new record. Because of how the availability list is ordered, this is the smallest hole that can hold the new record.

If the hole is larger than the new record, the left-over fragment is saved as a new hole at its sorted position in the availability list.

Hint. Use C's qsort() function to sort the availability list.

C.4.3 Worst Fit

If your program sees the availability order `--worst-fit`, it will order entries in the availability list sorted in descending order of hole size. New holes are inserted into the availability list in the proper sorted position. If multiple holes have the same size, the entries for these holes should be sorted in ascending order of hole offset.

When a new record is added, a worst-fit strategy is used to examine the first entry in the availability list to see if it is large enough to hold the new record. Because of how the availability list is ordered, this is the largest hole that can hold the new record.

If the hole is larger than the new record, the left-over fragment is saved as a new hole at its sorted position in the availability list.

Hint. Use C's qsort() function to sort the availability list.

C.5   USER INTERFACE

The user will communicate with your program through a set of commands typed at the keyboard. Your program must support four simple commands:

- add key rec
  Adds a new record rec with an SID of key to the student file. The format of rec is a |-delimited set of fields (exactly as described in the Student File section above), for example,

    add 712412913 712412913|Ford|Rob|Phi

  adds a new record with an SID of 712412913, a last of Ford, a first of Rob, and a major of Phi.

- find key
  Finds the record with SID of key in the student file, if it exists. The record should be printed in |-delimited format, (exactly as described in the Student File section above), for example,

    find 712412913

  should print on-screen

    712412913|Ford|Rob|Phi

- del key
  Delete the record with SID of key from the student file, if it exists.

- end
  End the program, close the student file, and write the index and availability lists to the corresponding index and availability files.

C.5.1   Add

To add a new record to the student file

1. Binary search the index for an entry with a key value equal to the new rec's SID. If such an entry exists, then rec has the same primary key as a record already in the student file. Write

    Record with SID=key exists

  on-screen, and ignore the add request, since this is not allowed. If the user wants to update an already existing record, they must first delete it, then re-add it.

2. Search the availability list for a hole that can hold rec plus the record size integer that precedes it.

  If a hole is found, remove it from the availability list, and write rec's size and body to the hole's offset. If the hole is bigger than rec plus its record size integer, there will be a fragment left at the end of the hole. Add the fragment back to the availability list as a new, smaller hole.

If no appropriately sized hole exists in the availability list, append rec's size and body to the end of the student file.

3. Regardless of where rec is written, a new entry must be added to the index holding rec's key and offset, maintaining the index in key-sorted order.

C.5.2   Find

To find a record, binary search the index for an entry with a key value of key. If an entry is found, use its offset to locate and read the record, then print the record on-screen. If no index entry with the given key exists, write

```
No record with SID=key exists
```

on-screen.

C.5.3   Del

To delete a record, binary search the index for an entry with a key value of key.

If an entry is found, use the entry's offset to locate and read the size of the record. Since the record is being deleted, it will form a hole in the student file. Add a new entry to the availability list containing the new hole's location and size. Remember, the size of the hole is the size of the record being deleted, plus the size of the integer preceding the record. Finally, remove the entry for the deleted record from the index. If no index entry with the given key exists, write

```
No record with SID=key exists
```

on-screen.

C.5.4   End

This command ends the program by closing the student file, and writing the index and availability lists to their corresponding index and availability list files.

C.6   PROGRAMMING ENVIRONMENT

All programs must be written in C, and compiled to run on a Linux system.

C.6.1 Writing Results

When your program ends, you must print the contents of your index and availability lists. For the index entries, print a line containing the text Index: followed by one line for each key–offset pair, using the following format.

```
printf( "key=%d: offset=%ld\n", index[i].key, index[i].off );
```

Next, for the availability list entries, print a line containing the text Availability: followed by one line for each size–offset pair, using the following format.

```
printf( "size=%d: offset=%ld\n", avail[i].siz, avail[i].off );
```

Finally, you must determine the number of holes hole_n in your availability list, and the total amount of space hole_siz occupied by all the holes (i.e., the sum of the sizes of each hole). These two values should be printed using the following format.

```
printf( "Number of holes: %d\n", hole_n );
printf( "Hole space: %d\n", hole_siz );
```

This will allow you to compare the efficiency of different availability list orderings to see whether they offer better or worse performance, in terms of the number of holes they create, and the amount of space they waste within the student file.

Your assignment will be run automatically, and the output it produces will be compared to known, correct output using diff. Because of this, **your output must conform to the above requirements exactly.** If it doesn't, diff will report your output as incorrect, and it will be marked accordingly.

C.7 SUPPLEMENTAL MATERIAL

In order to help you test your program, we provide two input and six output files. The input files contain commands for your program. You can use file redirection to pass them in as though their contents were typed at the keyboard.

- input-01.txt, an input file of commands applied to an initially empty student file (http://go.ncsu.edu/big-data-assn2-input-01.txt), and

- input-02.txt, an input file of commands applied to the student file produced by input-01.txt (http://go.ncsu.edu/big-data-assn2-input-02.txt).

The output files show what your program should print after each input file is processed.

- output-01-first.txt, the output your program should produce after it processes input-01.txt using --first-fit (http://go.ncsu.edu/big-data-assn2-output-01-first.txt),

- output-02-first.txt, the output your program should produce after it processes input-02.txt using --first-fit (http://go.ncsu.edu/big-data-assn2-output-02-first.txt),

- output-01-best.txt, the output your program should produce after it processes input-01.txt using --best-fit (http://go.ncsu.edu/big-data-assn2-output-01-best.txt),

- output-02-best.txt, the output your program should produce after it processes input-02.txt using --best-fit (http://go.ncsu.edu/big-data-assn2-output-02-best.txt),

- output-01-worst.txt, the output your program should produce after it processes input-01.txt using --worst-fit (http://go.ncsu.edu/big-data-assn2-output-01-worst.txt), and

- output-02-worst.txt, the output your program should produce after it processes input-02.txt using --worst-fit (http://go.ncsu.edu/big-data-assn2-output-02-worst.txt).

For example, to test your program using --best-fit, you would issue the following commands:

```
% rm student.db
% assn_2 --best-fit student.db < input-01.txt > my-output-01-best.txt
% assn_2 --best-fit student.db < input-02.txt > my-output-02-best.txt
```

Note. As shown in the example above, you start a "new" student database by removing any existing student file. If your program sees the student file doesn't exist, it can assume that the index and availability files shouldn't exist, either. You can handle this assumption in any way you want. One simple approach would be to open the index and availability files in w+b mode, which enables reading and writing, and automatically discards any existing files with the same names.

You can use diff to compare output from your program to our output files. If your program is running properly and your output is formatted correctly, your program should produce output identical to what is in these files.

Please remember, the files we're providing here are meant to serve as examples only. Apart from holding valid commands, and the previous guarantees of a limit of 10,000 key–offset and 10,000 size–offset pairs, **you cannot make any assumptions** about the content of the input files we will use to test your program.

C.8 HAND-IN REQUIREMENTS

Use the online assignment submission system to submit the following files:

- assn_2, a Linux executable of your finished assignment, and

- all associated source code files (these can be called anything you want).

There are four important requirements that your assignment must satisfy.

1. Your executable file must be named exactly as shown above. The program will be run and marked electronically using a script file, so using a different name means the executable will not be found, and subsequently will not be marked.

2. Your program must be compiled to run on a Linux system. If we cannot run your program, we will not be able to mark it, and we will be forced to assign you a grade of zero.

3. Your program must produce output that exactly matches the format described in the Writing Results section of this assignment. If it doesn't, it will not pass our automatic comparison to known, correct output.

4. You must submit your source code with your executable prior to the submission deadline. If you do not submit your source code, we cannot MOSS it to check for code similarity. Because of this, any assignment that does not include source code will be assigned a grade of zero.

Assignment 3: Mergesort

FIGURE D.1 Disk-based mergesort

T HE GOALS for this assignment are two-fold:

1. To introduce you to sorting data on disk using mergesort.

2. To compare the performance of different algorithms for creating and merging runs during mergesort.

D.1 INDEX FILE

During this assignment you will sort a binary index file of integer key values. The values are stored in the file in a random order. You will use a mergesort to produce a second index file whose key values are sorted in ascending order.

D.2 PROGRAM EXECUTION

Your program will be named `assn_3` and it will run from the command line. Three command line arguments will be specified: a mergesort method, the name of the input index file, and the name of the sorted output index file.

```
assn_3 mergesort-method index-file sorted-index-file
```

Your program must support three different mergesort methods.

1. `--basic`   Split the index file into sorted runs stored on disk, then merge the runs to produce a sorted index file.

2. `--multistep`   Split the index file into sorted runs. Merge subsets of runs to create super-runs, then merge the super-runs to produce a sorted index file.

3. `--replacement`   Split the index file into sorted runs created using replacement selection, then merge the runs to produce a sorted index file.

For example, executing your program as follows

```
assn_3 --best-fit student.db
```

would order holes in the availability list in ascending order of hole size, and would use `student.db` to store records.

Note. For convenience, we refer to the input index file as input.bin and the output sorted index file as sort.bin throughout the remainder of the assignment.

D.3 AVAILABLE MEMORY

Mergesort's run sizes and merge performance depend on the amount of memory available for run creating and merging runs.

Your program will be assigned one input buffer for reading data (e.g., blocks of keys during run creation of parts of runs during merging). The input buffer must be sized to hold a maximum of 1000 integer keys.

Your program will also be assigned one output buffer for writing data (e.g., sorted blocks of keys during run creation or sorted subsets of sort.bin during merging). The output buffer must be sized to hold a maximum of 1000 integer keys.

D.4 BASIC MERGESORT

If your program sees the merge method `--basic`, it will implement a standard mergesort of the keys in input.bin. The program should perform the following steps.

1. Open `input.bin` and read its contents in 1000-key blocks using the input buffer.

2. Sort each block and write it to disk as a run file. You can use any in-memory sorting algorithm you want (e.g., C's qsort() function). Name each run file index-file.*n*, where *n* is a 3-digit run identifier, starting at 0. For example, if the input index file is input.bin, the run files would be named

```
input.bin.000
input.bin.001
input.bin.002
...
```

3. Open each run file and buffer part of its contents into the input buffer. The amount of each run you can buffer will depend on how many runs you are merging (e.g., merging 50 runs using the 1000-key input buffer allows you to buffer 20 keys per run).

4. Merge the runs to produce sorted output. Use the output buffer to write the results in 1000-key chunks as binary data to sort.bin.

5. Whenever a run's buffer is exhausted, read another block from the run file. Continue until all run files are exhausted.

You must record how much time it takes to complete the basic mergesort. This includes run creation, merging, and writing the results to sort.bin.

Note. You will never be asked to merge more than 1000 runs in Step 3. This guarantees there will always be enough memory to assign a non-empty buffer to each run.

Note. Do not erase the intermediate run files. They will be needed during the grading of your assignment.

D.5 MULTISTEP MERGESORT

If your program sees the merge method --multistep, it will implement a two-step mergesort of the keys in input.bin. The program should perform the following steps.

1. Create the initial runs for input.bin, exactly like the basic mergesort.

2. Merge a set of 15 runs to produce a super-run. Open the first 15 run files and buffer them using your input buffer. Merge the 15 runs to produce sorted output, using your output buffer to write the results as binary data to a super-run file.

3. Continue merging sets of 15 runs until all of the runs have been processed. Name each super-run file index-file.super.*n*, where *n* is a 3-digit super-run

identifier, starting at 0. For example, if the input file is input.bin, the super-run files would be named

```
input.bin.super.000
input.bin.super.001
input.bin.super.002
  . . .
```

Note. If the number of runs created in Step 1 is not evenly divisible by 15, the final super-run will merge fewer than 15 runs.

4. Merge all of the super-runs to produce sorted output. Use the input buffer to read part of the contents of each super-run. Use the output buffer to write the results in 1000-key chunks as binary data to sort.bin.

   You must record how much time it takes to complete the multistep mergesort. This includes initial run creation, merging to create super-runs, merging super-runs, and writing the results to sort.bin.

Note. You will never be asked to merge more than 1000 super-runs in Step 3. This guarantees there will always be enough memory to assign a non-empty buffer to each super-run.

Note. Do not erase the intermediate run files. They will be needed during the grading of your assignment.

D.6 REPLACEMENT SELECTION MERGESORT

If your program sees the merge method --replacement, it will implement a merge-sort that uses replacement selection to create runs from the values in input.bin. The program should perform the following steps.

1. Divide your input buffer into two parts: 750 entries are reserved for a heap $H_1 \ldots H_{750}$, and 250 entries are reserved as an input buffer $B_1 \ldots B_{250}$ to read keys from input.bin.

2. Read the first 750 keys from input.bin into H, and the next 250 keys into B. Rearrange H so it forms an ascending heap.

3. Append H_1 (the smallest value in the heap) to the current run, managed through the output buffer. Use replacement selection to determine where to place B_1.

   - If $H_1 \leq B_1$, replace H_1 with B_1.
   - If $H_1 > B_1$, replace H_1 with H_{750}, reducing the size of the heap by one. Replace H_{750} with B_1, increasing the size of the secondary heap by one.

   Adjust H_1 to reform H into a heap.

4. Continue replacement selection until *H* is empty, at which point the current run is completed. The secondary heap will be full, so it replaces *H*, and a new run is started.

5. Run creation continues until all values in input.bin have been processed. Name the runs exactly as you did for the basic mergesort (i.e., input.bin.000, input.bin.001, and so on).

6. Merge the runs to produce sorted output, exactly like the merge step in the basic mergesort.

You must record how much time it takes to complete the replacement selection mergesort. This includes replacement selection run creation, merging the replacement selection runs, and writing the results to sort.bin.

Note. You will never be asked to merge more than 1000 runs in Step 6. This guarantees there will always be enough memory to assign a non-empty buffer to each run.

Note. Do not erase the intermediate run files. They will be needed during the grading of your assignment.

D.7 PROGRAMMING ENVIRONMENT

All programs must be written in C, and compiled to run on a Linux system.

D.7.1 Measuring Time

The simplest way to measure execution time is to use gettimeofday() to query the current time at appropriate locations in your program.

```
#include <sys/time.h>

struct timeval tm;

gettimeofday( &tm, NULL );
printf( "Seconds: %d\n", tm.tv_sec );
printf( "Microseconds: %d\n", tm.tv_usec );
```

Comparing tv_sec and tv_usec for two timeval structs will allow you to measure the amount of time that's passed between two gettimeofday() calls.

D.7.2 Writing Results

Sorted keys must be written to sort.bin as binary integers. C's built-in file writing operations allow this to be done very easily.

```
#include <stdio.h>

FILE *fp;                      /* Output file stream */
int    output_buf[ 1000 ];   /* Output buffer */

fp = fopen( "sort.bin", "wb" );
fwrite( output_buf, sizeof( int ), 1000, fp );
fclose( fp );
```

Your program must also print the total execution time for the mergesort it performs as a single line on-screen. Assuming the execution time is held in a timeval struct called exec_tm, use the following printf statement to do this.

```
printf( "Time: %ld.%06ld", exec_tm.tv_sec, exec_tm.tv_usec );
```

Your assignment will be run automatically, and the output it produces will be compared to known, correct output using diff. Because of this, your output must conform to the above requirements exactly. If it doesn't, diff will report your output as incorrect, and it will be marked accordingly.

D.8 SUPPLEMENTAL MATERIAL

In order to help you test your program, we provide example input.bin and sort.bin files.

- input.bin, a binary input file containing 250,000 keys (http://go.ncsu.edu/big-data-assn3-input.bin), and

- sort.bin, a binary output file containing input.bin's 250,000 keys in ascending sorted order (http://go.ncsu.edu/big-data-assn3-sort.bin).

You can use diff to compare output from your program to our sort.bin file. Please remember, the files we're providing here are meant to serve as examples only. Apart from holding integers, and a guarantee that the number of runs (or super-runs) will not exceed the input buffer's capacity, **you cannot make any assumptions** about the size or the content of the input and sort files we will use to test your program.

D.9 HAND-IN REQUIREMENTS

Use the online assignment submission system to submit the following files:

- assn_3, a Linux executable of your finished assignment, and

- all associated source code files (these can be called anything you want).

There are four important requirements that your assignment must satisfy.

1. Your executable file must be named exactly as shown above. The program will be run and marked electronically using a script file, so using a different name means the executable will not be found, and subsequently will not be marked.

2. Your program must be compiled to run on a Linux system. If we cannot run your program, we will not be able to mark it, and we will be forced to assign you a grade of zero.

3. Your program must produce output that exactly matches the format described in the Writing Results section of this assignment. If it doesn't, it will not pass our automatic comparison to known, correct output.

4. You must submit your source code with your executable prior to the submission deadline. If you do not submit your source code, we cannot MOSS it to check for code similarity. Because of this, any assignment that does not include source code will be assigned a grade of zero.

Assignment 4: B-Trees

FIGURE E.1 B-Trees

THE GOALS for this assignment are two-fold:

1. To introduce you to searching data on disk using B-trees.

2. To investigate how changing the order of a B-tree affects its performance.

E.1   INDEX FILE

During this assignment you will create, search, and manage a binary index file of integer key values. The values stored in the file will be specified by the user. You will structure the file as a B-tree.

E.2 PROGRAM EXECUTION

Your program will be named `assn_4` and it will run from the command line. Two command line arguments will be specified: the name of the index file, and a B-tree order.

```
assn_4 index-file order
```

For example, executing your program as follows

```
assn_4 index.bin 4
```

would open an index file called `index.bin` that holds integer keys stored in an order-4 B-tree. You can assume order will always be ≥ 3. For convenience, we refer to the index file as `index.bin` throughout the remainder of the assignment.

Note. If you are asked to open an existing index file, you can assume the B-tree order specified on the command line matches the order that was used when the index file was first created.

E.3 B-TREE NODES

Your program is allowed to hold individual B-tree nodes in memory—but not the entire tree—at any given time. Your B-tree node should have a structure and usage similar to the following.

```c
#include <stdlib.h>

int order = 4;       /* B-tree order */

typedef struct {     /* B-tree node */
    int    n;        /* Number of keys in node */
    int    *key;     /* Node's keys */
    long   *child;   /* Node's child subtree offsets */
} btree_node;

btree_node node;     /* Single B-tree node */

node.n = 0;
node.key = (int *) calloc( order - 1, sizeof( int ) );
node.child = (long *) calloc( order, sizeof( long ) );
```

Note. Be careful when you're reading and writing data structures with dynamically allocated memory. For example, trying to write node like this

```c
fwrite( &node, sizeof( btree_node ), 1, fp );
```

will write node's key count, the pointer value for its key array, and the pointer value for its child offset array, but **it will not** write the contents of the key and child offset arrays. The arrays' contents and not pointers to their contents need to be written explicitly instead.

```
fwrite( &node.n, sizeof( int ), 1, fp );
fwrite( node.key, sizeof( int ), order - 1, fp );
fwrite( node.child, sizeof( long ), order, fp );
```

Reading node structures from disk would use a similar strategy.

E.3.1   Root Node Offset

In order to manage any tree, you need to locate its root node. Initially the root node will be stored near the front of index.bin. If the root node splits, however, a new root will be appended to the end of index.bin. The root node's offset will be maintained persistently by storing it at the front of index.bin when the file is closed, and reading it when the file is opened.

```
#include <stdio.h>

FILE *fp;    /* Input file stream */
long  root;  /* Offset of B-tree root node */

fp = fopen( "index.bin", "r+b" );

/*  If file doesn't exist, set root offset to unknown and create
 *  file, otherwise read the root offset at the front of the file */

if ( fp == NULL ) {
   root = -1;
   fp = fopen( "index.bin", "w+b" );
   fwrite( &root, sizeof( long ), 1, fp );
} else {
   fread( &root, sizeof( long ), 1, fp );
}
```

E.4   USER INTERFACE

The user will communicate with your program through a set of commands typed at the keyboard. Your program must support four simple commands:

- add k
  Add a new integer key with value k to index.bin.

- find k
  Find an entry with a key value of k in index.bin, if it exists.

- print
  Print the contents and the structure of the B-tree on-screen.

- end

Update the root node's offset at the front of index.bin, close the index file, and end the program.

E.4.1 Add

Use a standard B-tree algorithm to add a new key k to the index file.

1. Search the B-tree for the leaf node L responsible for k. If k is stored in L's key list, print

   ```
   Entry with key=k already exists
   ```

   on-screen and stop, since duplicate keys are not allowed.

2. Create a new key list K that contains L's keys, plus k, sorted in ascending order.

3. If L's key list is not full, replace it with K, update L's child offsets, write L back to disk, and stop.

4. Otherwise, split K about its median key value k_m into left and right key lists $K_L = (k_0, \ldots, k_{m-1})$ and $K_R = (k_{m+1}, \ldots, k_{n-1})$. Use ceiling to calculate $m = \lceil (n-1)/2 \rceil$. For example, if $n = 3$, $m = 1$. If $n = 4$, $m = 2$.

5. Save K_L and its associated child offsets in L, then write L back to disk.

6. Save K_R and its associated child offsets in a new node R, then append R to the end of the index file.

7. Promote k_m, L's offset, and R's offset and insert them in L's parent node. If the parent's key list is full, recursively split its list and promote the median to its parent.

8. If a promotion is made to a root node with a full key list, split and create a new root node holding k_m and offsets to L and R.

E.4.2 Find

To find key value k in the index file, search the root node for k. If k is found, the search succeeds. Otherwise, determine the child subtree S that is responsible for k, then recursively search S. If k is found during the recursive search, print

```
Entry with key=k exists
```

on-screen. If at any point in the recursion S does not exist, print

```
Entry with key=k does not exist
```

on-screen.

E.4.3 Print

This command prints the contents of the B-tree on-screen, level by level. Begin by considering a single B-tree node. To print the contents of the node on-screen, print its key values separated by commas.

```
int         i;      /* Loop counter */
btree_node  node;   /* Node to print */
long        off;    /* Node's offset */

for( i = 0; i < node.n - 1; i++ ) {
  printf( "%d,", node.key[ i ] );
}
printf( "%d", node.key[ node.n - 1 ] );
```

To print the entire tree, start by printing the root node. Next, print the root node's children on a new line, separating each child node's output by a space character. Then, print their children on a new line, and so on until all the nodes in the tree are printed. This approach prints the nodes on each level of the B-tree left-to-right on a common line.

For example, inserting the integers 1 through 13 inclusive into an order-4 B-tree would produce the following output.

```
1: 9
2: 3,6 12
3: 1,2 4,5 7,8 10,11 13
```

Hint. To process nodes left-to-right level-by-level, do not use recursion. Instead, create a queue containing the root node's offset. Remove the offset at the front of the queue (initially the root's offset) and read the corresponding node from disk. Append the node's non-empty subtree offsets to the end of the queue, then print the node's key values. Continue until the queue is empty.

E.4.4 End

This command ends the program by writing the root node's offset to the front of index.bin, then closing the index file.

E.5 PROGRAMMING ENVIRONMENT

All programs must be written in C, and compiled to run on a Linux system.

Your assignment will be run automatically, and the output it produces will be compared to known, correct output using diff. Because of this, **your output must conform to the print command's description**. If it doesn't, diff will report your output as incorrect, and it will be marked accordingly.

E.6   SUPPLEMENTAL MATERIAL

In order to help you test your program, we provide example input and output files.

- input-01.txt, an input file of commands applied to an initially empty index file saved as an order-4 B-tree (http://go.ncsu.edu/big-data-assn4-input-01.txt), and

- input-02.txt, an input file of commands applied to the index file produced by input-01.txt (http://go.ncsu.edu/big-data-assn4-input-02.txt).

The output files show what your program should print after each input file is processed.

- output-01.txt, the output your program should produce after it processes input-01.txt (http://go.ncsu.edu/big-data-assn4-output-01.txt), and

- output-02.txt, the output your program should produce after it processes input-02.txt (http://go.ncsu.edu/big-data-assn4-output-02.txt).

You can use diff to compare output from your program to our output files. If your program is running properly and your output is formatted correctly, your program should produce output identical to what is in these files.

Please remember, the files we're providing here are meant to serve as examples only. Apart from holding valid commands, **you cannot make any assumptions** about the size or the content of the input files we will use to test your program.

E.7   HAND-IN REQUIREMENTS

Use the online assignment submission system to submit the following files:

- assn_4, a Linux executable of your finished assignment, and

- all associated source code files (these can be called anything you want).

There are four important requirements that your assignment must satisfy.

1. Your executable file must be named exactly as shown above. The program will be run and marked electronically using a script file, so using a different name means the executable will not be found, and subsequently will not be marked.

2. Your program must be compiled to run on a Linux system. If we cannot run your program, we will not be able to mark it, and we will be forced to assign you a grade of zero.

3. Your program must produce output that exactly matches the format described in the Writing Results section of this assignment. If it doesn't, it will not pass our automatic comparison to known, correct output.

4. You must submit your source code with your executable prior to the submission deadline. If you do not submit your source code, we cannot MOSS it to check for code similarity. Because of this, any assignment that does not include source code will be assigned a grade of zero.

Index